MB122

AF497495

MICROBIOLOGY

(PAPER-II)
MICROBIAL CULTIVATION & GROWTH

[2 Credits]

For
F.Y.B.Sc. (Semester – II)
As Per New Revised Syllabus, CBCS Pattern
From June 2019

Dr. Pragati Sunil Abhyankar
M.Sc., Ph.D., SET
Associate Professor, Department of Microbiology,
Haribhai V. Desai College of Arts, Science and Commerce
Pune 411 002.

Dr. Rajashree Bhalchandra Patwardhan
M.Sc., M.Phil, Ph.D., (Microbiology),
Associate Professor, Department of Microbiology,
Haribhai V. Desai College of Arts, Science and Commerce
Pune 411 002.

N5043

Microbiology (Paper - II) ISBN 978-93-89686-93-7

First Edition : January 2020
© : Authors

Published By:
NIRALI PRAKASHAN
Abhyudaya Pragati, 1312, Shivaji Nagar
Off J.M. Road, PUNE – 411005
Tel - (020) 25512336/37/39, Fax - (020) 25511379
Email : niralipune@pragationline.com

➢ DISTRIBUTION CENTRES

PUNE

Nirali Prakashan
(For orders within Pune) : 119, Budhwar Peth, Jogeshwari Mandir Lane, Pune 411002, Maharashtra, Tel : (020) 2445 2044, Mobile : 9657703145
Email : niralilocal@pragationline.com

Nirali Prakashan
(For orders outside Pune) : S. No. 28/27, Dhayari, Near Asian College Pune 411041
Tel : (020) 24690204; Mobile : 9657703143
Email : bookorder@pragationline.com

MUMBAI

Nirali Prakashan : 385, S.V.P. Road, Rasdhara Co-op. Hsg. Society Ltd., Girgaum, Mumbai 400004, Maharashtra;
Mobile : 9320129587 Tel : (022) 2385 6339 / 2386 9976, Fax : (022) 2386 9976
Email : niralimumbai@pragationline.com

➢ DISTRIBUTION BRANCHES

JALGAON

Nirali Prakashan : 34, V. V. Golani Market, Navi Peth, Jalgaon 425001, Maharashtra, Tel : (0257) 222 0395, Mob : 94234 91860;
Email : niralijalgaon@pragationline.com

KOLHAPUR

Nirali Prakashan : New Mahadvar Road, Kedar Plaza, 1st Floor Opp. IDBI Bank, Kolhapur 416 012, Maharashtra. Mob : 9850046155;
Email : niralikolhapur@pragationline.com

NAGPUR

Nirali Prakashan : Above Maratha Mandir, Shop No. 3, First Floor, Rani Jhanshi Square, Sitabuldi, Nagpur 440012, Maharashtra
Tel : (0712) 254 7129;
Email : niralinagpur@pragationline.com

DELHI

Nirali Prakashan : 4593/15, Basement, Agarwal Lane, Ansari Road, Daryaganj Near Times of India Building, New Delhi 110002
Mob : 08505972553, Email : niralidelhi@pragationline.com

BENGALURU

Nirali Prakashan : Maitri Ground Floor, Jaya Apartments, No. 99, 6th Cross, 6th Main, Malleswaram, Bengaluru 560003, Karnataka;
Mob : 9449043034
Email: niralibangalore@pragationline.com

Other Branches : Hyderabad, Chennai

niralipune@pragationline.com | www.pragationline.com

Also find us on www.facebook.com/niralibooks

उसका उपयोग करते हैं। वे रूप (रूपा), ध्वनि (सबदा), गंध (गंध), स्वाद (रस) और स्पर्श (स्पर्श) हैं।

सांस और इंद्रियों के बीच संबंध

कुछ पहले के उपनिषदों ने इंद्रियों को श्वास या श्वास (प्राण) के पहलुओं के रूप में भी वर्णित किया है। वे पाँच श्वासों से भिन्न हैं, जिनका उल्लेख उनमें किया गया है, अर्थात् प्राण, व्यान, समान, उदान और अपान, जिस दिशा में वे शरीर में प्रवाहित होते हैं, उसके अनुसार नाम दिए गए हैं। हालांकि, इंद्रियों को उनके साथ समान किया जाता है, क्योंकि श्वास को शरीर के सभी अंगों का स्वामी माना जाता है।

उपनिषदों में भी इसके बारे में दो रूपक हैं। श्वास इन्द्रियों का स्वामी है क्योंकि वह कामनाओं के प्रति अभेद्य है। अंग बुरी इच्छाओं की घेरे में हैं, जबकि सांस अनैच्छिक है और उनके अधीन नहीं है। अतः श्वास अग्नि के समान शुद्ध करने वाला है और जिस प्रकार स्वर्ग में अग्नि शरीर को अर्पित किए गए भोजन को स्वीकार करता है और अंगों में उनके उचित हिस्से के अनुसार वितरित करता है। शरीर जीवित है, और अंग तब तक सक्रिय है, जब तक उनमें श्वास का संचार होता है। इसलिए वह भी उनका सहारा है। जब किसी व्यक्ति की मृत्यु होती है, तो सांसों के साथ-साथ अंगों में मौजूद देवताओं ने शरीर को छोड़ दिया और मध्य क्षेत्र की यात्रा की, जहां से वे अपने-अपने क्षेत्रों में लौट जाते हैं।

नाम दिए गए हैं। हालांकि, इंद्रियों को उनके साथ समान किया जाता है, क्योंकि श्वास को शरीर के सभी अंगों का स्वामी माना जाता है।

उपनिषदों में भी इसके बारे में दो रूपक हैं। श्वास इन्द्रियों का स्वामी है क्योंकि वह कामनाओं के प्रति अभेद्य है। अंग बुरी इच्छाओं की चपेट में हैं, जबकि सांस अनैच्छिक है और उनके अधीन नहीं है। अतः श्वास अग्नि के समान शुद्ध करने वाला है और जिस प्रकार स्वर्ग में अग्नि शरीर को अर्पित किए गए भोजन को स्वीकार करता है और अंगों में उनके उचित हिस्से के अनुसार वितरित करता है। शरीर जीवित है, और अंग तब तक सक्रिय हैं, जब तक उनमें श्वास का संचार होता है। इसलिए वह भी उनका सहारा है। जब किसी व्यक्ति की मृत्यु होती है, तो सांसों के साथ-साथ अंगों में मौजूद देवताओं ने शरीर को छोड़ दिया और मध्य क्षेत्र की यात्रा की, जहाँ से वे अपने-अपने क्षेत्रों में लौट आए।

Preface ...

We are extremely pleased to present this book **F.Y.B.Sc. Microbiology (Paper-II) - Microbial Cultivation and Growth** to all the students who have opted for the subject with great interest. The syllabi for F.Y.B.Sc. Microbiology have been revised and modified so as to widen the scope of the subject to be compatible to present developments and needs of the subject. Our effort is to provide the students with the best guidelines in order to help them to achieve the expected outcomes in these changed circumstances. This book covers the entire new and revised syllabus for the first semester of F.Y.B.Sc. (Microbiology) as prescribed by SPPU. The present book includes the following:

- o A complete coverage of the topics for semester-II
- o Simple scientific language for easy understanding of the subject
- o Precise presentation
- o Topic wise explanation
- o Questions at the end of each topic
- o Use of elaborative and appropriate diagrams

We are confident that, this will be extremely helpful for the students to approach the new revised syllabus and give them confidence to successfully face the revised and reformed examination pattern.

We express a deep sense of gratitude to our publishers Mr. Dineshbhai Furia and Mr. Jigneshbhai Furia of Nirali Prakashan for having shown the confidence in us for accomplishment of the task. We appreciate Dr. G. S. Gugale, our colleague for having boosted us to take up this work and moreover to take it to completion. We are also thankful to our colleague, Dr. Mrs. Suniti Gore who readily gave her expert suggestions while finalizing the book. Our sincere thanks also to all those who directly and indirectly became a part of this.

We have given our best inputs for this book. Any suggestions towards the improvement of this book and sincere comments are most welcome on niralipune@pragationline.com.

With best wishes to all the students who are always our strength.

PUNE   **AUTHORS**

January 2020

Syllabus ...

Credit I: Cultivation of Microorganisms   (15 Hrs.)

1. Nutritional requirements and nutritional classification
2. Design and preparation of media: Common ingredients of media and types of media
3. Methods for cultivating photosynthetic, extremophilic and chemo-lithotrophic bacteria, anaerobic bacteria, algae, fungi, actinomycetes and viruses
4. Concept of Enrichment, Pure Culture, Isolation of culture by streak plate, pour plate, spread plate
5. Maintenance of bacterial and fungal cultures using different techniques
6. Culture collection centres and their role
7. Requirements and guidelines of National Biodiversity Authority for culture collection centres

Credit II: Bacterial Growth   (15 Hrs.)

1. Kinetics of bacterial growth (Exponential growth model)
2. Growth curve and Generation time
3. Diauxic growth
4. Measurement of bacterial growth- Methods of enumeration:
    - Microscopic methods (Direct microscopic count, counting cells using improved Neubauer, Petroff-Hausser's chamber)
    - Plate counts (Total viable count)
    - Turbidometric methods (including Nephelometry)
    - Estimation of biomass (Dry mass, Packed cell volume)
    - Chemical methods (Cell carbon and nitrogen estimation)
5. Factors affecting bacterial growth {pH, Temperature, Solute Concentration (Salt and Sugar)} and Heavy metals

Contents ...

Chapter **1**...

Cultivation of Microorganisms

Learning Objectives...

- ➢ To understand nutritional requirements of microorganisms.
- ➢ To understand nutritional types of bacteria.
- ➢ To understand physical conditions required for growth.
- ➢ To understand the choice of media and conditions of incubation.
- ➢ To understand about the maintenances of microbial cultures.

Heinrich Hermann Robert Koch

Heinrich Hermann Robert Koch was a German physician, microbiologist and one of the main founders of modern bacteriology. Koch's early research in this laboratory yielded one of his major contributions to the field of microbiology, as he developed the technique of growing bacteria. Furthermore, he managed to isolate and cultivate selected pathogenic bacteria in pure laboratory culture. In an attempt to cultivate bacteria, Koch began to use solid nutrients such as potato slices.

Through these initial experiments, Koch observed individual colonies of identical, pure cells. He found that potato slices were not suitable media for all organisms, and later began to use nutrient solutions with gelatin. However, he soon realized that gelatin, like potato slices, was not the optimal medium for bacterial growth, as it did not remain solid at 37°C, the ideal temperature for growth of most human pathogens. As suggested to him by Walther and Fanny Hesse, Koch began to utilize agar to grow and isolate pure cultures, because this polysaccharide remains solid at 37°C, is not degraded by most bacteria, and results in a transparent medium.

1.1 NUTRITIONAL REQUIREMENTS

All living forms including microorganisms require food for their growth. Microorganisms must get all the substances from the environment required for the synthesis of their cell materials and for the generation of energy. These substances are termed as **Nutrients**. These nutrients are processed by the cell through metabolic activities into the various structural and functional cell organelles. A culture medium must contain all the nutrients in appropriate quantities to fulfil the specific requirements of microorganisms. The microorganisms are extremely diverse in their physiological properties and correspondingly in their specific nutrient requirements. The chemical composition of microbial cell remains constant throughout the life. Water constitutes about 80 to 90% of the total weight of cells and therefore, it is a major essential nutrient required for the microbial cells. The solid matter of cells contains in addition to hydrogen and oxygen, carbon, nitrogen, phosphorous and sulphur in order of decreasing requirement. These six elements constitute 95% cellular dry weight. Other elements like potassium, magnesium, calcium, iron, manganese, cobalt, copper, molybdenum and zinc are required in trace amount for all the microorganisms. All the required metallic elements can be supplied as nutrients in the form of cations of inorganic salts.

Table 1.1: Elemental composition of microbial cell and its physiological function

Sr. No	Element	% cell dry weight	Physiological function
1.	Hydrogen	8	Component of cellular water, organic cell materials
2.	Oxygen	20	Component of cellular water, organic cell materials, as O_2, electron acceptor in aerobes respiration
3.	Carbon	50	Component of organic cell materials
4.	Nitrogen	14	Component of proteins, coenzymes, nucleic acids
5.	Sulphur	1	Component of proteins: amino acids cysteine and methionine, Coenzyme A and cocarboxylase

Contd...

Sr. No	Element	% cell dry weight	Physiological function
6.	Phosphorus	3	Component of nucleic acids, phospholipids and coenzymes
7.	Potassium	1	Cofactor of some enzymes, inorganic cation in the cell
8.	Sodium	1	Helps in transport (sodium pump), cation in the cell
9.	Magnesium	0.5	Cation in the cell, inorganic cofactor for many enzymatic reactions involving ATP, constituent of chlorophyll, functions in binding enzymes to substrates
10.	Manganese	Trace	Inorganic cofactor for some enzymes
11.	Calcium	0.5	Cation in the cell, cofactor for some enzymes
12.	Iron	0.2	Constituent of cytochrome and heme and non-heme proteins
13.	Cobalt	Trace	Constituent of Vitamin B_{12} and its coenzyme derivative
14.	Copper, Zink, Nickel, Molybdenum	Trace	Inorganic constituents of special enzymes

Some microorganisms have additional specific mineral requirements. Diatoms and some algae synthesize cell walls that are saturated with silica so have silicon requirement which is supplied as silicate. Marine bacteria like *Cyanobacteria* and photosynthetic bacteria require high concentration of sodium. Chemolithotrophs like *Thiobacillus* require high concentration of sulphur and iron which can be supplied as elemental sulphur and ferrous sulphate. Many other essential elements are involved in the transport of materials across the cell membrane. Sodium is required for permease enzyme activity. Some elements ate required as cofactors which assist the enzymes in their activity. Fe^{++} is required by the cytochromes, dehydrogenases and catalase.

The Role of Different Elements:

1. Carbon:

Microorganisms carrying out photosynthesis and organisms which obtain energy from oxidation of inorganic compounds use the oxidised form of carbon, CO_2 as the only source of cellular carbon. Conversion of CO_2 to organic cell constituents through a reductive process requires energy which is derived from light or from oxidation of reduced inorganic compounds. In non-photosynthetic organisms carbon is obtained from organic nutrients. These organic substrates are at the same oxidation level as the organic cell constituents and so they don't require to undergo reduction to serve as carbon source. Thus, most of the carbon containing organic substrates directly enter the energy yielding metabolic pathways and finally the carbon is excreted from the cell as CO_2 (major end product of energy yielding respiratory metabolism) or the mixture of CO_2 and organic compounds (end products of fermentative metabolism). Organic substrates thus have a dual nutritional role; both as carbon and energy source. So many microorganisms use a single organic compound to supply both these nutritional requirements. Other microorganisms require variable number of other organic compounds as nutrients having purely biosynthetic function. These organic compounds are termed as **growth factors** which are required as precursors of certain organic cell components, which the microorganism is not able to synthesize. Microorganisms are tremendously diverse with respect to type and number of organic compounds which they use as primary source of carbon and energy. Any naturally produced organic compound can serve as a source of carbon and energy by some microorganism. *Pseudomonas* can use any one of 90 different organic compounds as their sole carbon and energy source while some methane oxidising bacteria can use only methane and methanol, and cellulose decomposing bacteria can use only cellulose as their only carbon and energy source.

2. Nitrogen and Sulphur:

Nitrogen and sulphur occur in the organic compounds of the cell in the reduced form as amino ($R-NH_2$) and sulfhydryl ($R-SH$) groups, respectively. Most photosynthetic organisms and many non-photosynthetic bacteria and fungi assimilate these elements in their oxidised inorganic state, as nitrates and sulphates. Nitrogen and sulphur requirements can be fulfilled by protein degradation products

Table 1.2: Functions of the Growth Factors

Sr. No.	Growth factor	Function
1.	Amino acids	Constituent of cellular proteins.
2.	Purines and pyrimidines	Constituent of nucleic acids.
3.	Vitamins	Organic compounds that form parts of prosthetic groups or active centres of some enzymes.
	i. Nicotinic acid (Niacin)	Coenzyme pyridine nucleotides NAD^+ and $NADP^+$. In dehydrogenation reactions
	ii. Riboflavin (Vitamin B_2)	Coenzyme Flavin nucleotides FAD and FMN In dehydrogenation reactions and electron transport
	iii. Thiamine (Vitamin B_1)	Coenzyme thiamine pyrophosphate Carboxylations and group transfer reactions
	iv. Pyridoxine (Vitamin B_6)	Coenzyme Pyridoxyl phosphate Amino acid metabolism, Transamination, Deamination, Decarboxylation reactions
	v. Pantothenic acid	Coenzyme A Keto acid oxidation and fatty acid metabolism
	vi. Folic acid	Coenzyme tetrahydrofolic acid Transfer of one carbon units
	vii. Biotin	Coenzyme Biotin CO_2 fixation, Carboxyl transfer
	viii. Cobalamin (Vitamin B_{12})	Molecular rearrangement reactions

like amino acids or peptone. These compounds also provide organic carbon and energy source along with nitrogen and sulphur. Requirement of reduced sulphur is met by providing sulphide or an organic compound containing sulfhydryl group like cysteine or mercapto compounds. Some nitrogen fixing microorganisms utilize atmospheric dinitrogen.

3. Growth Factors:

Any organic compound that an organism requires as a precursor or component of its organic cell material, which it cannot synthesize from simpler carbon sources and must be provided as a nutrient is known as growth factor. These include amino acids, purines, pyrimidines and vitamins.

Growth factors are required in very small amounts just to fulfil the specific needs in the biosynthetic reactions. They can be provided naturally as pure vitamins or amino acids or artificially as yeast extract or beef extract. Some organisms require growth factors in the nutrient medium without which they cannot grow. These organisms are called as **auxotrophs**. Organisms termed **fastidious** tend to require a variety of growth factors. Those organisms which can synthesize their own growth factors through their biosynthetic pathways are termed as **prototrophs**.

4. Oxygen:

It is a universal component of cell and is provided in large amounts in the form of water. Many microorganisms require molecular oxygen (O_2). These organisms are dependent on aerobic respiration for their energy requirements and for which molecular oxygen functions as terminal oxidizing agent. These organisms are called **obligate aerobes.** Some microorganisms obtain energy through the reactions that do not involve utilization of molecular oxygen. For these organisms molecular oxygen is a toxic substance which either kills them or inhibits their growth. These organisms are called **obligate anaerobes.** Some microorganisms grow in presence or in absence of molecular oxygen. Some organisms like lactic acid bacteria have fermentative energy yielding metabolism and are unresponsive to the presence of oxygen. These are termed as **aerotolerant anaerobes**. Other microorganisms like enteric bacteria, yeasts can shift from respiratory to fermentative mode of metabolism.

These are called **facultative anaerobes** which use O_2 as terminal oxidizing agent when it is available but can also get energy in its absence by fermentative reactions. Some obligate aerobes grow well at partial pressures of oxygen, 0.2 atm; below that is present in air. These are known as **microaerophilic.** These organisms comprise the enzymes which get inactivated under strongly oxidising conditions and remain in their functional state at low partial pressures of O_2. Some microaerophilic bacteria which obtain energy by the oxidation of molecular hydrogen produce hydrogenase enzyme involved in hydrogen utilization. These bacteria get inactivated by oxygen.

1.2 NUTRITIONAL CLASSIFICATION (CLASSIFICATION OF MICROORGANISMS BASED ON THEIR MODE OF NUTRITION)

Table 1.3

Sr. No.	Nutritional type	Source of carbon and energy	Examples of organisms
1.	Chemoautotrophs	Inorganic chemical substances as source of energy and CO_2 as main source of carbon	Nitrifying hydrogen, sulphur and iron bacteria
2.	Chemoheterotrophs	Organic chemical substances as source of energy and organic compounds as source of carbon	Most bacteria, fungi and protozoa
3.	Photoautotrophs	Light as a source of energy and CO_2 as main source of carbon	Purple and green sulphur bacteria, algae, cyanobacteria
4.	Photoheterotrophs	Light as a source of energy and organic compounds as source of carbon	Purple and green non sulphur bacteria

Microorganisms are broadly classified in two main nutritional classes. The most useful simple nutritional classification is based on two parameters: The nature of energy source and nature of principle carbon source.

1. **Phototrophs:** Organisms that depend on light as a source of energy.

2. **Chemotrophs:** Organisms that use chemical compounds as a source of energy.

Autotrophs are the organisms which use CO_2 (inorganic source) as principal carbon source while heterotrophs are the organisms which obtain carbon from organic nutrients.

1.3 DESIGN AND PREPARATION OF MEDIA: COMMON INGREDIENTS OF MEDIA AND TYPES OF MEDIA

In making a culture medium for any microorganism, the main goal is to provide a balanced mixture of the essential nutrients, at concentrations that will permit good growth. The nutrient medium must fulfil the functions for which it is designed like differentiation, growth, enrichment, selection etc. Many organic constituents like sugars or salts of fatty acids like acetate as well as some inorganic constituents like phosphates (in case of algae) become growth inhibitory or toxic at the high concentration.

Culture media are of fundamental importance to obtain pure cultures, to grow and count microbial cells, and to cultivate and select microorganisms. A microbiological culture medium encourages the growth and survival of microorganisms. While studying microorganisms in the laboratory we require a single cell i.e. a pure culture. To obtain a pure culture, the microbial cells have to be isolated by using several methods like pour plating and streak plating. Specific culture media minimise the steps required for the isolation of desired organism.

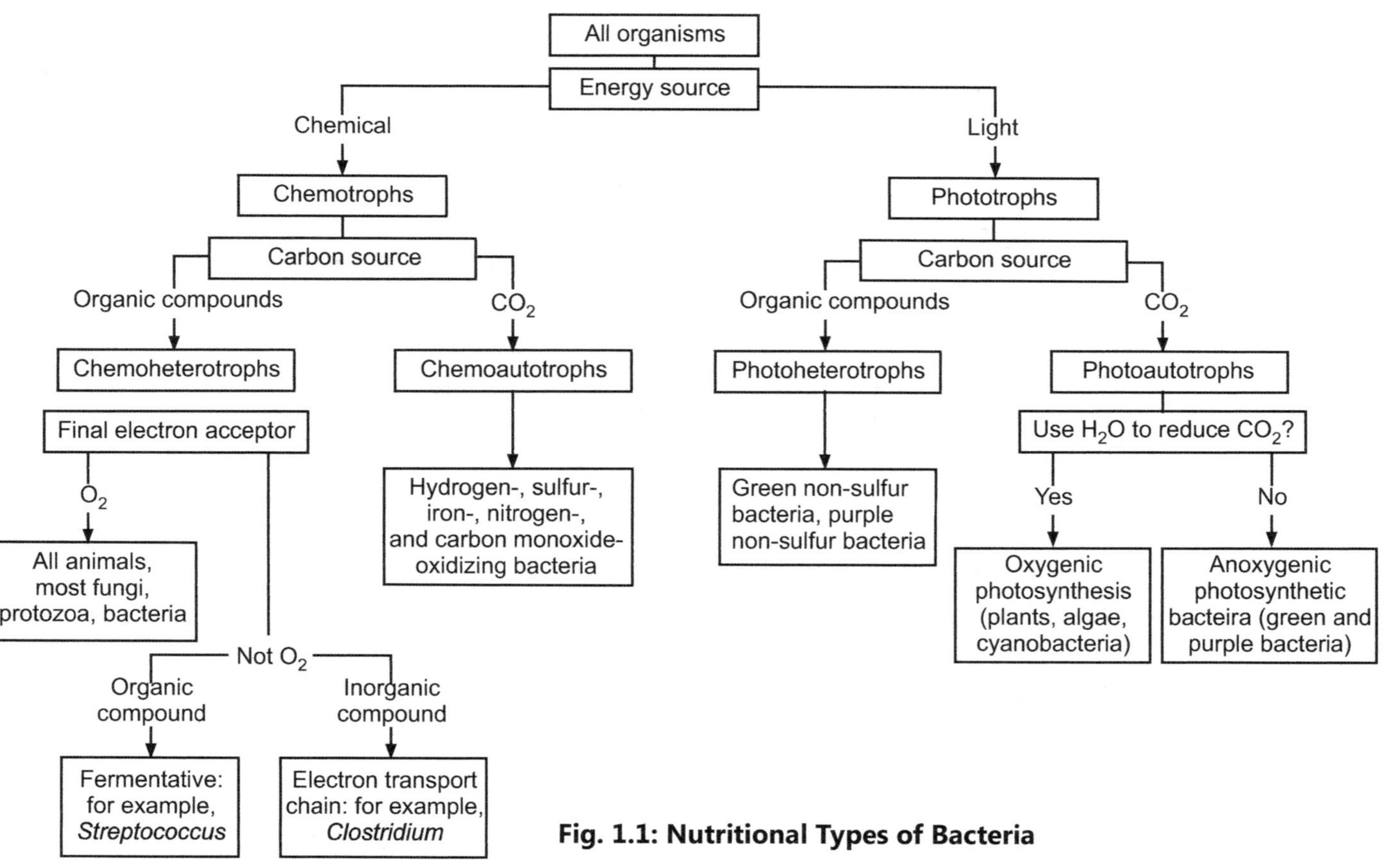

Fig. 1.1: Nutritional Types of Bacteria

Pure cultures (population of cells arising from a single cell) of microorganisms are required for

a. The identification of species
b. Large scale production of metabolites
c. Study of molecular make up of different types of bacteria
d. Preparation of consortium

1.3.1 Culture Media are Divided in Two Categories

1. Synthetic media: They are composed of compounds of known chemical composition. They contain all the ingredients as inorganic salts or most of the ingredients as inorganic salts and only one or two complex but chemically defined organic substances. Synthetic media can be exactly duplicated in every batch. These media are used to precisely determine the nutritional requirements of bacteria.

Table 1.4

Common ingredients	Additional ingredients			
	Medium 1	**Medium 2**	**Medium 3**	**Medium 4**
Water 1 liter K_2HPO_4 1 gm $MgSO_4.7H_2O$ 200 mg $FeSO_4.7H_2O$ 10 mg $CaCl_2$ 10 mg Trace elements Mn, Mo, Cu, Co, Zn, 0.02 – 0.5 mg of each	NH_4Cl 1 gm	Glucose 5 gm NH_4Cl 1 gm	Glucose 5 gm NH_4Cl 1 gm Nicotinic acid 0.1 mg	Glucose 5 gm Yeast extract 5 mg
	Chemo-autotrophs, Nitrifying bacteria Nitro-somonas	Anaerobes, fungi	*Proteus vulgaris*	Aerobic and anaerobic chemo-autotrophs

2. Non-synthetic media: They contain ingredients of unknown chemical composition like yeast extract, beef or meet extract, peptone, blood, serum etc. It is not possible to exactly duplicate the composition of non-synthetic medium, even if the ingredients are used in the same amounts. Most of the media used in the microbiology laboratory are of non-synthetic type.

When a medium is designed for any organism, it must be understood as to what degree of complexity is required by the growth of the organism. The following table 1.4 shows four media of increasing complexity.

Glucose must be sterilized separately and added aseptically. When sugars are heated in presence of other ingredients like phosphates, they are partially decomposed to substances that are toxic to some microorganisms.

1.3.2 Some of the Environmental Parameters which are Related to the Culture Medium

1. pH:

For the growth, the organisms require optimum pH range at which they can replicate at a fast rate. So at all times during the growth of the organism, the optimum pH of the growth medium should be maintained. Some organisms during their metabolic activity produce acidic or alkaline metabolites which changes the pH of medium. Such acidic or alkaline conditions can arrest the growth of the organism growing in the medium. To maintain the constant pH buffers or insoluble carbonates are added in the medium during its preparation. They can prevent the change in pH by neutralizing acid or alkali component as it is formed. Phosphate buffers consist of mixtures of monohydrogen and dihydrogen phosphates (K_2HPO_4 - slightly basic and KH_2PO_4 - weakly acidic). Equimolar solutions of K_2HPO_4 and KH_2PO_4 result in neutral solution (pH 6.8). When any change in hydrogen ion concentration (production of acidic products) takes place in culture medium, part of salt is converted to a weak acid as follows.

$$K_2HPO_4 + HCl \longrightarrow KH_2PO_4 + KCl$$

When the reaction becomes alkaline, the opposite reaction occurs.

$$KH_2PO_4 + KOH \longrightarrow K_2HPO_4 + H_2O$$

Therefore, such solutions act as buffers and resist changes in pH of the medium.

Uses of balanced C and N sources have the capacity to buffer the medium with the use of proteins, peptides, and amino acids. pH can also be controlled by addition of NAOH or NH_3 and H_2SO_4.

2. Temperature:

Increase in temperature increase the enzyme activity. But if temperature becomes too high, enzyme activity diminishes and the protein (the enzyme) denatures. Every bacterial species has specific growth temperature requirements which are largely determined by the temperature requirements of its enzymes. Very high and very low temperatures obstruct the enzyme processes microorganisms depend on to survive. Each enzyme has a specific temperature range in which its activity is highest. In the cultivation of microorganisms temperature control is provided by the use of incubators maintained by thermostatic device at the desired value.

3. Oxygen Concentration:

Microorganisms vary in their requirements for molecular oxygen. Obligate aerobes depend on aerobic respiration and use oxygen as a terminal electron acceptor. Oxygen is essential nutrient for obligately aerobic bacteria. Facultative anaerobes show better growth in the presence of oxygen but will also grow without it. Aerobic microorganisms can be grown easily on the surface of agar plates and in shallow layers of liquid medium. Various types of shaking machines which constantly agitate and aerate the medium are available for laboratory use. The air may be introduced through a porous 'sparger' in the medium which delivers oxygen in the form of very fine bubbles.

For anaerobic microorganisms, the atmosphere must be oxygen free. To eliminate the oxygen, the culture media are placed within containers where carbon dioxide and hydrogen gas are generated, and oxygen is removed from the atmosphere. Commercially available products achieve these conditions. Anaerobic chambers (McFielde's jar) can also be used within closed compartments, and technicians can manipulate culture media within these chambers. To encourage carbon dioxide formation, a candle can be burned to use up oxygen and replace it with carbon dioxide.

4. Provision of Light:

In case of photoautotophic and photoheterotrophic organisms, light serves as a source of energy. Hence, laboratory cultures of such microorganisms are illuminated naturally or artificially for obtaining their growth. Direct exposure to sunlight is avoided because the intensity becomes too high and so the temperature rises due to which the growth of organisms is inhibited. The emission spectrum of the lamp employed is important. Fluorescent light sources produce relatively less heat; therefore maintenance of suitable temperature is not difficult. They are suitable for the cultivation of algae and cyanobacteria which perform photosynthesis with light of wavelength shorter than 700 nm. Incandescent lamps with light of 750 – 1000 nm wavelength are used for the cultivation of purple and green bacteria. A light cabinet can be made with internal incandescent illumination where the temperature can be controlled by ventilation or refrigeration.

1.3.3 Common Ingredients of Media

1. Peptone: Peptones are excellent natural sources of amino acids, peptides and proteins in growth media. They are most often obtained by enzymatic digestion or acid hydrolysis of natural products, such as animal tissues, milk, plants or microbial cultures. The enzymes used for hydrolysis are proteolytic enzymes like papain or trypsin. The hydrolytic activity leads to formation of proteoses, peptones, peptides and amino acids. The function of peptone is to serve as a source of nitrogen. It also provides carbon. Since, it is proteinaceous and amphoteric in nature it also acts as buffer.

2. Meat extract: Meat or beef extract is manufactured from meat with low fat and considered to complement the nutritive properties of peptone by contributing minerals, phosphates, energy sources and those essential growth factors missing from peptone. It provides amino acid, vitamins and mineral salts (phosphate and sulphate). The meat pieces are extracted by boiling. The extract is dried under vacuum till the paste is obtained. The paste is dried to flakes and powdered. It provides creatinine, xanthine, hypoxanthine, uric acid, adenylic acid, inosinic acid, carnosine, carnitine, beta alanine, glutamine and urea. The non-nitrogenous extractives include glycogen, hexose, phosphates, lactic acid, succinic acid, inositol, fat and inorganic salts. It is a source of vitamins of B complex group as thiamine, riboflavin and pyridoxine. Other vitamins include pantothenic acid and nicotinic acid, biotin, folic acid, beta alanine, glutamine and para amino benzoic acid.

3.　Yeast extract: It is used to stimulate the growth of bacteria. It provides nitrogenous compounds, carbon, sulfur, trace nutrients, vitamin B complex and other important growth factors, which are essential for the growth of diverse microorganisms. It is rich in vitamins, minerals, and digested nucleic acids. Yeast extract is obtained from spent yeast from distilleries or breweries. The first step of yeast extract production is fermentation. Sugar is added to nourish the yeast. In addition, a temperature of 30°C and an enough supply of oxygen are fed to large fermenter which allows growth of yeasts. Then the yeast is concentrated and washed in centrifuges to remove the residual sugar. To extract the cells, they are heated to 45°C in distilled water and the pH of the suspension is adjusted to 6.5. The yeast stops growing at approximately 40°C. The suspension is stirred for 16 hours. Heating is controlled to avoid denaturation of the contents on autolysis. Due to stirring cell contents come into the suspension. It is dried under vacuum to obtain the powder.

4.　Sodium chloride: The presence of sodium chloride in growth medium maintains a salt concentration in the medium that is similar to the cytoplasm of the microorganisms. If the salt concentration is not similar, osmosis takes place transporting excess water into or out from the cell. Sodium chloride is an essential component in complex, synthetic and semisynthetic media. It maintains isotonicity of the medium required for appropriate transport of nutrients into the cell.

5.　Mineral salts: Microbes require very small amounts of mineral elements, such as iron, copper, molybdenum, and zinc; which are referred to as trace elements. Most are essential for activity of certain enzymes, usually as cofactors. Phosphate salts of potassium and sodium are added as buffers, ammonium salts serve as nitrogen source. Ferrous sulphate is added as source of Fe^{++}. Traces of magnesium, sodium, potassium, calcium, chlorides are also ingredients of many microbial media. The amount of mineral salts required is very less; between 0.02 to 0.1 mg/lit.

6. Agar: Agar-agar is a jelly-like substance, inert polysaccharide, obtained from seaweed or marine red algae belonging to genus Gelidium (*G. corneum* and related species). These algae are found growing in the water off the coasts of Southern California, Sri Lanka, Malaya and Japan. Agar is a mixture of two components: the linear polysaccharide agarose, and a heterogeneous mixture of smaller molecules called **agaropectin**. It is the sulfuric acid ester of a linear galactan. It is insoluble in water. It forms the supporting structure in the cell walls of certain species of algae and is released on boiling. These algae are known as **agarophytes** and belong to the Rhodophyta (red algae) phylum. It is a major solidifying agent used in bacteriological media at concentration of 2 – 3 % (w/v), dissolves at 90-100°C, solidify at 45°C. It is usually added after the other medium components have been added and dissolved in the water. Once solidified, the medium will not melt until brought back up to about 100°C. Some species of bacteria; Pseudomonas, Achromobacter, *Vibrio* and *Cytophaga* can digest agar.

7. Water: It is very essential for bacterial growth. Water is necessary for life processes, and hence is very important in culture media. Water from different sources varies in composition. Hard and tap water containing calcium and magnesium can react with phosphates, contained in ingredients such as peptone and beef extract, and precipitates are formed. For such reasons, distilled water is preferred. Distilled water also needs to be used with care because metal-distilled and glass-distilled water can vary in the pH values. Tap water can be used in some cases where autotrophs are to be grown.

8. Carbohydrates: They provide bacteria with energy and carbon source.

Monosaccharides: Glucose, Galactose Arabinose, Xylose

Disaccharides: Lactose, Sucrose, Maltose

Trisaccharides: Raffinose

Polysaccharides: Starch, Cellulose, Glycogen, Dextrin,

Polyhydric alcohols: Glycerol, Mannitol, Sorbitol, Ribitol,

Glucosides: Salicin

Non-Carbohydrates: Inositol

1.3.4 Types of Media

1. Based on nature of Ingredients:

a. Living Media: The media that contain living cells. These are used for cultivation of viruses, rickettsia and chlamydia etc. These organisms are obligate intracellular parasites and needs living cells to grow.

    i. Embryonated chicken eggs: The fertile chicken eggs when incubated for 5 to 12 days can be used for the cultivation of intracellular parasites. They are inoculated in the embryonic tissues. This technique is used in the production of vaccines against smallpox, yellow fever, influenza and other diseases. It is the most economical and convenient method for the cultivation of animal viruses.

    ii. Tissue culture: Tissue culture is the growth of tissues or cells in an artificial medium separate from the organism. This is typically facilitated via use of a liquid, semi-solid, or solid growth medium, such as broth or agar. Tissue culture commonly refers to the culture of animal cells and tissues, with the more specific term plant tissue culture being used for plants. The term "tissue culture" was coined by American pathologist Montrose Thomas Burrows. In 1907, Ross Harrison for the first-time developed tissue culture from frog tissue.

    iii. Use of laboratory animals: Some viruses like HIV cannot be cultivated in tissue culture or in embryonated chicken eggs. They are cultivated in living animals like mice, rabbit, guinea pigs.

b. Non-living Media: The media that contain non-living substances like chemicals or natural ingredients. These are divided in three different types.

    i. Natural media: These media are prepared from natural ingredients like milk, blood, serum, vegetable juices e.g. Tomato juice or Carrot juice. These media were widely used during olden days of microbiology. Few are still in use. They are not prepared based on exact composition. They contain organic and inorganic nutrients which support growth of most bacteria/fungi. They are cheaper and convenient to use but are not reproducible.

ii. **Semisynthetic media:** They contain both, natural ingredients of unknown composition and pure chemicals of known composition. Natural ingredients used are yeast, beef. meat extracts, tryptone, peptone, blood, serum etc. Examples of media includes Nutrient broth/agar, MacConkey's agar, Sabouraud's agar. They provide satisfactory growth results of the desired organism because of complex composition but they are not reproducible.

iii. **Synthetic media:** They contain the pure chemical ingredients of known exact composition. Therefore, they are reproducible. These are of two types:

   a. **Inorganic media:** They contain inorganic ingredients of known concentration. They are used for cultivation of autotrophs.

   b. **Organic media:** They contain organic ingredients of known concentration. They are used for cultivation of heterotrophs. Synthetic media are particularly used in research and analytical studies. Because of use of pure chemicals, they are expensive.

Minimal medium for the growth of *Bacillus megaterium*. An example of a chemically defined medium for growth of a heterotrophic bacterium.

Table 1.5

Component	Amount	Function of component
1. Sucrose	10.0 g	C and energy source
2. K_2HPO_4	2.5 g	pH buffer; P and K source
3. KH_2PO_4	2.5 g	pH buffer; P and K source
4. $(NH_4)_2HPO_4$	1.0 g	pH buffer; N and P source
5. $MgSO_4\ 7H_2O$	0.20 g	S and Mg^{++} source
6. $FeSO_4\ 7H_2O$	0.01 g	Fe^{++} source
7. $MnSO_4\ 7H_2O$	0.007 g	Mn^{++} Source
8. Water pH 7.0	985 ml	

2. Based on Function or Application:

(a) Selective media:

A selective media are composed of specific ingredients to inhibit the growth of certain species of microbes in a mixed culture while allowing others to grow. Selective media are used for enrichment of a single or group of organisms from a mixed population. The **liquid media** selects the organisms which grow fast among the various members of diverse population. Along with chemical composition (nutrients), other factors like temperature, pH, ionic strength, illumination, aeration and source of inoculum play an important role in selectivity of medium. Thermophiles (high temperature tolerating) can be isolated by selecting specific temperature. *Bacillus* or *Clostridium*, endospore producing organisms are grown in the medium by giving heat shock (80°C for 5 minutes) due to which only endospore containing organisms survive and can be isolated easily. When the number of organisms in the sample is higher, the selective media in solid form are used. The sample is streaked or spread on a solid medium. Due to less quantity of inoculum spatial distribution of the organism allows it to grow in the areas where competition for the nutrition is less and individual colonies can be isolated.

Examples of Selective Media include:

1. Thayer martin agar is used to select *Neisseria gonorrhoeae* which contains antibiotics; vancomycin, colistin and nystatin.

2. Potassium tellurite medium is used to recover *Corynebacterium diphtheriae* which contains 0.04% potassium tellurite.

3. MacConkey's agar is used for Enterobacteriaceae members which contains bile salt that inhibits most gram-positive bacteria. MacConkey agar allows the growth of intestinal organisms like *E. coli, Klebsiella, Shigella* since it contains sodium taurocholate as a selective agent. The intestinal organisms are adapted to sodium taurocholate in bile.

4. Pseudosel agar (Cetrimide Agar) is used to recover *P. aeruginosa* which contains cetrimide (antiseptic agent).

5. Crystal violet blood agar is used to recover *S. pyogenes* which contains 0.0002% crystal violet.

6.   Lowenstein Jensen medium used to recover *M. tuberculosis* is made selective by incorporating malachite green.

7.   Wilson and Blairs medium contains bismuth sulphite and brilliant green which permits growth of *Salmonella* species and inhibits coli forms.

8.   Selective media such as TCBS agar are used for isolating *V. cholerae* from faecal specimens have elevated pH (8.5-8.6), which inhibits most other bacteria.

9.   Mannitol salt agar contains 7.5 – 10% salt as a selective agent which inhibits other organisms and allows the growth of salt tolerating *Staphylococcus aureus*.

(b) Enriched media:

Some organisms are fastidious regarding their nutritional requirements. Enriched media contain the nutrients required to support the growth of fastidious ones. They are commonly used to harvest as many different types of microbes as are present in the specimen. Such a medium provides specific nutrients that encourage selected species of microorganisms to flourish in a mixed sample. To isolate *Salmonella* species from faecal samples, it is helpful to place a sample of the material in an enriched medium to encourage *Salmonella* species to multiply before the isolation techniques begin. Human pathogens require blood or serum components in the medium. Blood agar is an enriched medium in which nutritionally rich whole blood supplements the basic nutrients. It is necessary to grow and detect haemolytic activity of organisms like *Streptococcus pyogenes*. Lowenstein-Jensen's (LJ) medium contains egg albumin which serves a rich protein source and helps in the gelling of the medium. Chocolate agar is enriched with heat-treated blood (40–45 °C), which turns brown and gives the medium the colour for which it is named. Milk agar is used to grow lactic acid bacteria while egg yolk agar is used for *Bacillus* with lecithinase activity.

(c) Enrichment media:

These types of media are used for the isolation of organism from a mixed culture or natural habitat. Enrichment medium provides conditions in the medium so that only the desired organism can grow or at least outgrows remaining organisms. They are usually liquid in nature. They are used to increase the number of organisms present in less number or slow growing. An enrichment medium contains some component that permits the growth of specific types or species of bacteria, because they alone can utilize the component from their environment. It provides an environment, both chemical and physical that allows growth of desired organism. Like selective medium inhibitory agents are not used in this medium. Several serial transfers are required in enrichment medium after which the desired organism becomes predominant in the population.

An enrichment medium for non-symbiotic nitrogen-fixing bacteria omits a source of added nitrogen to the medium. The medium is inoculated with a potential source of these bacteria (e.g. a soil sample) and incubated in the atmosphere wherein the only source of nitrogen available is N_2. For free living nitrogen fixers like *Azotobacter*, Ashby's broth which lacks a nitrogen source is inoculated with soil sample. Species which can fix atmospheric nitrogen can grow in the medium and form the pellicle at the surface of the medium and hence can be isolated easily. Succinate broth is used for the enrichment of the purple non-sulphur photosynthetic bacteria. In this medium, a particular nutrient utilized by the desired organisms is included as the only carbon source. Most other organisms do not metabolize succinate under the anaerobic conditions utilized. Selenite Broth or tetrathionate broth is used as a selective enrichment medium for the cultivation of *Salmonella* spp. that may be present in small numbers and competing with intestinal flora. An enrichment medium, such as Loffler's medium, is used to grow *Corynebacterium diphtheriae*. A selective enrichment medium for growth of the extreme halophile (*Halococcus*) contains nearly 25% salt [NaCl], which is required by the extreme halophiles and which inhibits the growth of all other prokaryotes.

(d) Differential media:

They contain the specific ingredients which differentiates the growth of one kind of organism from other types by visual observation. Differential media are used to distinguish between closely related organisms or groups of organisms.

Table 1.6: Enrichment medium for growth of extreme halophiles

Component	Amount	Function of Component
1. Casamino acids	7.5 g	Source of amino acids, N, S and P
2. Yeast extract	10.0 g	Source of growth factors
3. Trisodium citrate	3.0 g	C and energy source
4. KCl	2.0 g	K^+ source
5. $MgSO_4$ 7 H_2O	20.0 g	S and Mg^{++} source
6. $FeCl_2$	0.023 g	Fe^{++} source
7. NaCl	250 g	Na^+ source for halophiles and inhibitory to non-halophiles
8. Water	1000 ml	pH 7.4

Examples of differential media include:

1. **MacConkey's Agar:** Organisms which ferment the lactose in the medium, lower the pH due to the production of acids. The pH indicator (neutral red) turns red, and the colonies become red. Other colonies on the same plate which do not contain lactose-fermenting cells appear colourless. It is a **selective-differential medium.** *E. coli* and *Klebsiella* are lactose fermenters. Salmonella, Shigella and Proteus are non-lactose fermenters. They do not produce acid. So, colonies of these organisms remain colourless.

2. **MacConkey's Agar:**
   Peptone   20.0 gm
   Sodium taurocholate   5.0 gm
   Agar   20.0 gm
   Neutral red   4 ml
   Lactose   10.0 gm
   Distilled water   1 litre

3. **Glucose Fermentation Broth.** This medium is used in tubes, with Durham tubes. Organisms which ferment the sugar (glucose) will cause the pH indicator to change colour upon production of acidic products. If insoluble gas (H_2) is produced during fermentation, bubbles are observed in the inverted Durham tube.

4. **Blood Agar** is a differential medium that distinguishes bacterial species by their ability to break down red blood cells. This medium differentiates between haemolytic and non-haemolytic bacteria.

5. **Mannitol Salt Agar:** It differentiates *Staphylococcus aureus* and *Staphylococus epidermidis.* The medium contains mannitol as a source of carbon and energy. The pH indicator used is phenol red. It is red at pH 8.4 and yellow at pH 6.8. *Staphylococcus aureus* ferments mannitol and produces acid hence colonies become yellow.

6. **Violet Red Bile Agar** is used to distinguish coliform bacteria such as *Escherichia coli* from no coliform organisms. The coliform bacteria appear as bright pink colonies in this media, while no coliforms appear a light pink or clear.

1.4 METHODS FOR CULTIVATING PHOTOSYNTHETIC, EXTREMOPHILIC AND CHEMO-LITHOTROPHIC

1.4.1 Methods for Cultivating Photosynthetic Bacteria

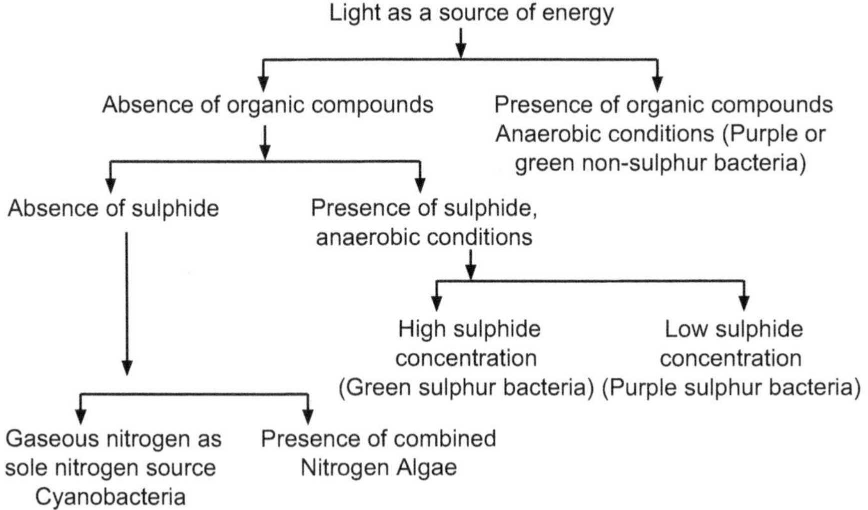

Fig. 1.2

The Winogradsky column is a simple device for culturing a large diversity of autotrophic microorganisms. Invented in the 1880s by Sergei Winogradsky, the device is a column of pond mud and water mixed with a carbon source such as newspaper (containing cellulose), blackened marshmallows or egg-shells (containing calcium carbonate), and a sulphur source such as gypsum (calcium sulphate) or egg yolk. Incubating the column in sunlight for month's results in an aerobic/anaerobic gradient as well as a sulphide gradient.

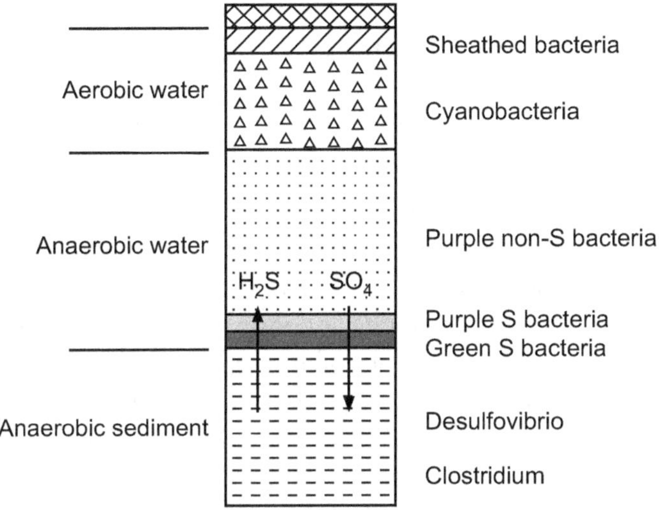

Fig. 1.3: Winogradoky column

The column (a glass tube, about 30 cm tall and 5 cm diameter) is filled with mud placed at the bottom to form 2-3 cm layer with shredded paper, calcium carbonate sulphate on the top of it. The column is then filled with a natural water sample from river or any water stream to 75% of total height of the jar. This provides numerous gradients, depending on additive nutrients, from which the variety of organisms can grow. The aerobic water phase and anaerobic mud or soil phase are one such distinction. Because of oxygen's low solubility in water, the water quickly becomes anoxic towards the crossing point of the mud and water. Anaerobic phototrophs are present to a large extent in the mud phase. Phototrophic algae, Cyanobacteria and other aerobic phototrophs are present along the surface and in the top layer of the column where there is enough air and presence of light. Organisms like green and purple sulphur bacteria in the middle layer and anaerobes like sulphate reducers e.g. *Desulfovibrio* grow in the bottom in the anaerobic layer.

1. **Purple sulphur bacteria:** These bacteria have the pigment bacteriochlorophyll located on the intracytoplasmic membrane i.e., thylakoids. These bacteria obtain energy from sulfur compounds. e.g. *Thiospirillum, Chromatium, Thiocystis, Thiocapsa, Theopedia*

2. **Purple non-sulphur bacteria:** Purple non-sulfur bacteria are anaerobic, facultative or obligate phototrophs. The column turns dark brown to bright red, or sometimes green, because of the organism's photopigments. e.g. *Rhodomicrobium, Rhodopseudomonas, Rhodospirillum, Rhodobacter.*

3. **Green sulphur bacteria:** These bacteria use hydrogen sulphide (H_2S) as hydrogen donor. The reaction takes place in the presence of light and pigment termed as bacteriovirdin or bacteriopheophytin or chlorobium chlorophyll. These bacteria take hydrogen from inorganic sources like sulphides and thiosulphates. Therefore, these bacteria are also known as photolithographs.

    e.g., *Chlorobium limicola, Chlorobacterium, Prosthecochloris* etc.

4. **Green non-sulphur bacteria:** Formerly known as gliding filamentous bacteria capable of anoxygenic photosynthesis. The green non-sulfur bacteria (phylum Chloroflexi) now are also known to comprise numerous chemotrophic bacteria of diverse ecophysiology and phylogeny. The most conspicuous representatives are the green- or orange-coloured thermophilic bacteria which form dense microbial mats in hot springs. They are typical photoorganoheterotrophs and well adapted to their changing environment by their gliding motility, tactic responses and versatile physiology.

    e.g., *Chloronema, Chloroflexux, Heliothrix, Roseiflexus, Oscillochloris*

Pringsheim's medium for the cultivation of Cyanobacteria / Blue Green Algae:

Potassium nitrate and ammonium hydrogen phosphate in the medium provide nitrogen source. Ferric chloride provides an iron source to blue green algae. Magnesium sulphate and the chloride salts are sources of ions that simulate metabolism.

Ingredients	**Gramm / Litre**
KNO_3 Potassium nitrate	0.200
$MgSO_4. 7H_2O$ Magnesium sulphate	0.010
Ammonium hydrogen phosphate	0.020
Calcium chloride	0.005
Iron (II) chloride	0.0005

1.4.2 Cultivation of Extremophiles

An extremophile (from Latin *extremus* meaning "extreme" and Greek *philia* meaning "love") is an organism with optimal growth in environmental conditions considered extreme in comparison to the environmental conditions that are comfortable to humans. Life can exist at extreme environments like the depth of the oceans, hot springs, the Arctic and the Antarctic, in very dry places, deep inside Earth, in harsh chemical environments, inside rocks and in high radiation environments etc. They thrive in habitats which for other terrestrial life-forms are intolerably hostile or even lethal. They thrive in extreme hot niches, ice, and salt solutions, as well as acid and alkaline conditions; some may grow in toxic waste, organic solvents, heavy metals, or in several other habitats that were previously considered inhospitable for life. Extremophiles teach us about the limits of life. They give us hope that life could exist on other planets and moons in this solar system and beyond.

Based on environmental conditions extremophiles are classified as shown in table 1.7:

Table 1.7

Sr. No.	Environmental condition	Type of organisms
1.	High temperature	Thermophiles
2.	Low temperature	Psychrophiles
3.	High pH	Alkaliphiles
4.	Low pH	Acidophiles
5.	High pressure	Barophiles
6.	High salt concentration	Halophiles
7.	High radiations	Radoduric/Radioresistant
8.	In presence of heavy metals	Metallotolerant/Heavy metal resistant

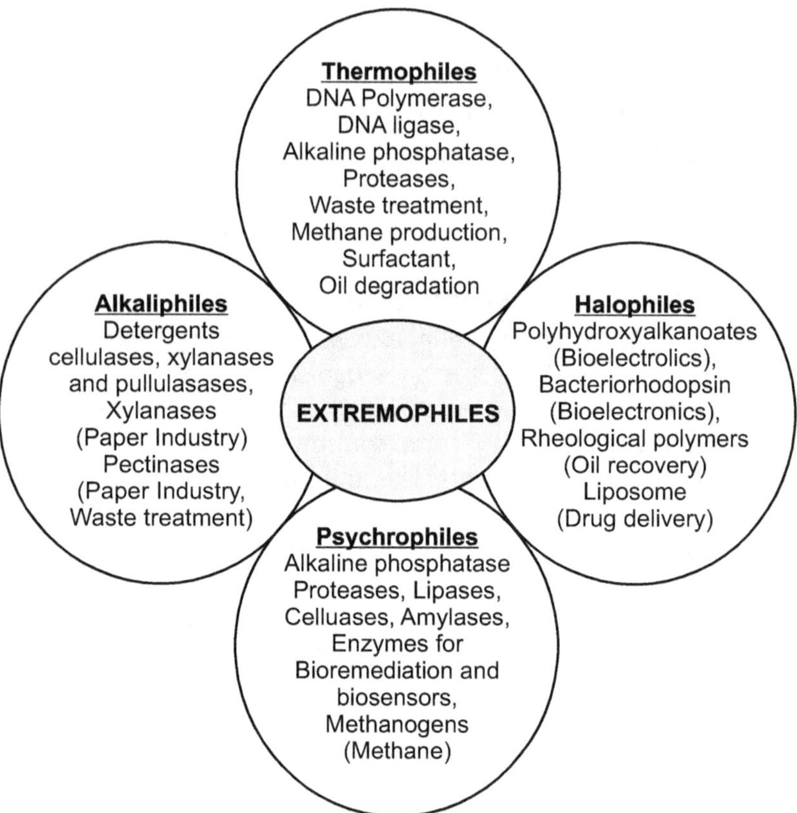

Fig. 1.4

Table 1.8

Type of Organisms	Minimum Temperature	Optimum Temperature	Maximum Temperature
Thermophilic	25 - 45	55	60 – 90
Mesophilic	5 - 25	18 - 45	30 -50
Psychrophilic	-20	15	30

Biotechnological applications of extremophiles in basic science, medicine and industry: Reagents such as Taq DNA polymerase, proteases etc. are important tools in biomedical research. Commercial exploitation of extremophiles in waste treatment, methane production, bioremediation, oil degradation etc. and their use in drug delivery, biosensors and paper industry are some of the applications.

Classification based on Temperature:

1. Thermophiles: "Thermophile" is derived from the Greek word *thermotita* meaning heat, and *philia* means love. A thermophile is an organism, a type of extremophile that thrives at relatively high temperatures, between 41 and 122°C (106 and 252°F). Many thermophiles are archaea. Thermophilic eubacteria are suggested to be the earliest bacteria on the earth. Thermophiles are found in various geothermally heated regions of the Earth, such as erupting volcanoes (1000°C), superheated hot springs (93 – 10°C) like those in Yellow stone National Park and deep sea hydrothermal vents, as well as decaying plant matter such as peat bogs and compost, non-boiling hot springs, superheated substrates like soils, rocks etc.

Table 1.9

Sr. No.	Bacteria	Optimum temperature	Maximum temperature
1.	*Bacillus stearothermophilus*	50 - 65	70 -75
2.	Photosynthetic bacteria Chlorobiaceae family	55	70 -73
3.	*Clostridium thermosaccharolyticum*	55	67
4.	*Sulfolobus acidocaldarius*	70 -75	85 - 90
5.	*Thermoactinomyces*		
6.	*Methanobacterium thermoautotropicum*	65 - 70	75
7.	*Thermus aquaticus*	70	79

Thermophiles are grown on complex media and incubated at high temperature depending on their optimum growth requirements.

Proteolytic enzymes produced by thermophiles are of considerable interest because they are stable and active at elevated temperatures. Moreover, they are resistant to organic solvents, detergents, low and high pH and other denaturants. Such properties allow many technological processes. The advantage of the use of thermostable enzymes for conducting biotechnological processes at elevated temperatures are: reducing the risk of contamination by mesophilic microorganisms; decreasing the viscosity of the reaction medium; increasing the bioavailability and solubility of organic compounds; increasing the diffusion coefficient of substrates and products resulting in higher reaction rates. Several cellulases, xylanases and pectinases have been applied in biotechnological processes such as biobleaching of paper pulp, production of animal feed, production of fermentable sugars for obtaining biofuel from cellulosic wastes, fruit juice extraction and clarification, refinement of vegetable fibres, degumming of natural fibres, curing of coffee, cocoa and tobacco and also for waste-water treatment.

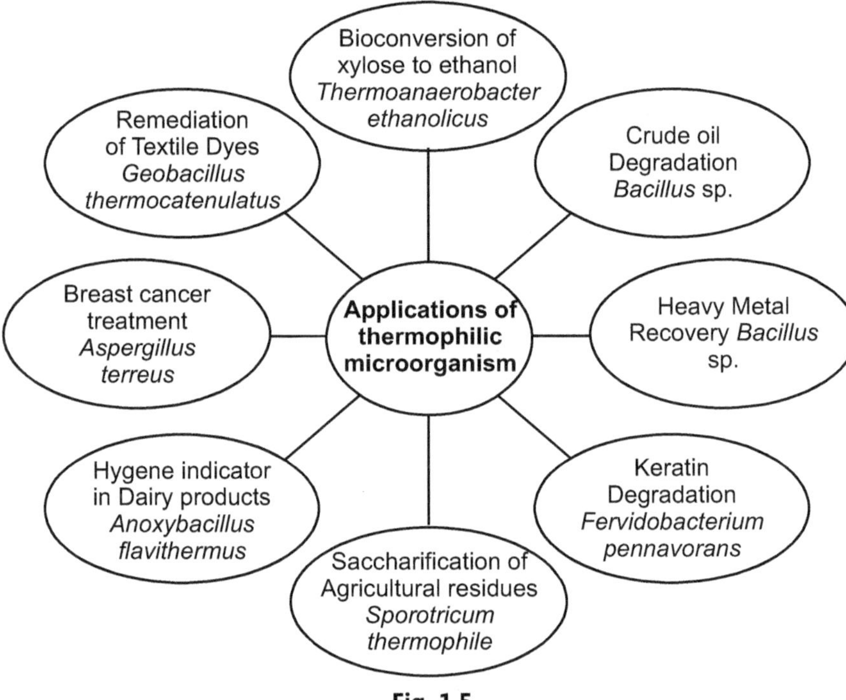

Fig. 1.5

2. Psychrophiles: **Psychrophiles** or cryophiles (adj. **Psychrophilic** or cryophilic) are extremophilic organisms that are capable of growth and reproduction in low temperatures, ranging from −20°C to +10°C. They are found in places that are permanently cold, such as the polar regions and the deep sea. Psychrophiles include bacteria, lichens, fungi, and insects. Among the bacteria that can tolerate extreme cold are spore and non-spore forming Gram positive rods, actinomycetes, Gram negative rods like *Pseudomonas, Arthrobacter, Flavobacterium, Psychrobacter* and members of the genera *Halomonas, Hyphomonas,* and *Sphingomonas. Chryseobacterium greenlandensis,* a psychrophile was found in 120,000-year-old ice. *Penicillium* is a genus of fungi found in a wide range of environments including extreme cold. *Umbilicaria Antarctica* and *Xanthoria elegans* are lichens that have been recorded photosynthesizing at temperatures ranging down to −24°C, and they can grow down to around −10°C. Organisms isolated from Antarctic valley were *Arthrobacter, Cellulomonas, Corynebacterium, Bacillus, Micrococcus* etc. Microorganisms can be classified as psychrotolerant that can tolerate low temperature and psychrophilic that requires low temperature for the growth. Cold-adapted microorganisms can be used as cell factories to produce unstable compounds as well as for bioremediation of polluted cold soils and waste waters.

3. Alkaliphiles/Basophiles:

Alkaliphiles are a class of extremophilic microbes capable of survival in alkaline (pH roughly 8.5–11) environments, growing optimally around a pH of 10. Alkaliphiles include prokaryotes, eukaryotes, and archaea. They are further categorized as obligate alkaliphiles (those that require high pH to survive), facultative alkaliphiles (those able to survive in high pH, but also grow under normal conditions) and haloalkaliphiles (those that require high salt content to survive). Haloalkaliphiles have been mainly found in extremely alkaline saline environments, such as alkaline soil, alkaline lakes e.g. Lonar lake in India, the Rift Valley lakes of East Africa and the western soda lakes of the United States. Many different kinds of alkaliphilic microorganisms, including bacteria belonging to the genera *Bacillus, Micrococcus, Pseudomonas,* and *Streptomyces* and eukaryotes such as yeasts and filamentous fungi, have been isolated from a variety of environments. *Nitrosomonas Nitrobacter* can survive at pH 13. Bacillus pasteurii, *Bacillus sphericus, Bacillus alkalophiles,* and *Bacillus circulans* can grow at pH 11. Penicillium, Fusarium, some protozoans like *Euglena gracilis* are alkalotolerant *Rhizobium, Agrobacterium* are acid tolerant. Other examples of alkaliphiles include *Halorhodospira halochloris, Natronomonas pharaonis,*

Thiohalospira alkaliphile. Alkaliphiles are grown in media whose pH has been adjusted in the alkaline range.

Alkaliphiles produce extracellular enzymes, which are able to function in their catalytic activities under high alkaline pH values because of their stability under these conditions. Proteases, protein degrading enzymes, are one of the most produced enzymes in industry. Among proteases, alkaline proteases, which are added to some detergents, are the most produced. Other alkaline enzymes, e.g. alkaline cellulases, alkaline amylases, and alkaline lipases, are also adjuncts to detergents for improving cleaning efficiency.

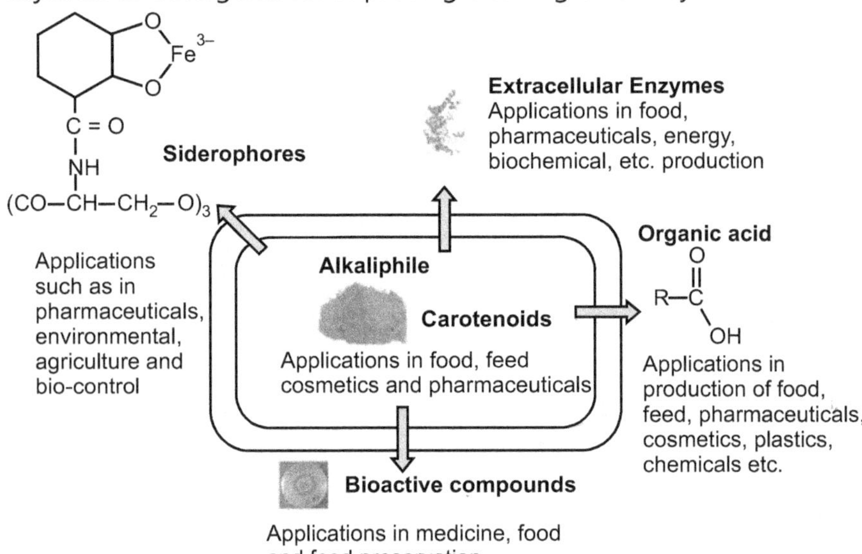

Fig. 1.6

4. Acidophiles:

These organisms thrive under highly acidic conditions (usually at pH 2.0 or below). They are found in different branches of the tree of life, including Archaea, Bacteria, fungi and Eukarya. *Acetobacter acidophilum* has an optimum pH of 3.0, lower limit of pH 2.8 and upper limit of 4.3. They are found in drainage waters and mining effluents. Acidity is due to production of sulfuric acid as a result of oxidation of sulphides. Acidophilic organisms are grown in liquid media at pH 2 - 3. At this acidic pH agar cannot solidify and therefore solid media are not used for isolation of acidophiles.

Examples of Acidophiles:

1. **Archaea:** *Sulfolobales, Thermoplasmatales, Halarchaeum acidiphilum,* facultatively anaerobic thermoacidophilic archaebacteria like *Acidianus brierleyi* and *A. infernus.*

2. **Bacteria:** *Bacillus, Micrococcus, Sarcina, Acidobacteria* e.g. *A. ferrooxidans, A. thiooxidans, Thiobacillus thioxidans, T. ferroxidans,* T. *prosperus, T. acidophilus, T. organovorus, T. cuprinus, Acetobacter aceti,* a bacterium that produces acetic acid (vinegar) from the oxidation of ethanol, *Alicyclobacillus,* a genus of bacteria that can contaminate fruit juices. Helicobacter pylori, a bacterium found in the human stomach, *Sulfolobus acidocaldarius* isolated from acid hot springs. (optimum pH range 2-3), *Thermoplasma acidophilum* which is cell wall less organism, *Saccharomyces.*

3. **Fungi:** *Aspergillus, Penicillium, Fusarium.*

4. **Eukarya:** *Mucor racemosus, Urotricha, Dunaliella acidophila.*

   Acid stable enzymes produced by acidophiles have applications in several industries such as starch, baking, fruit juice processing, animal feed and pharmaceuticals. Acidophiles are widely used in bioleaching of metals from low grade ores.

Table 1.10: Classification of Halophiles based on Salt Tolerance

Sr. No.	Halophiles	Salt concentration
1.	Slight halophiles *Erythrobacter flavus*	0.2 -0.5 M 1.7 to 4.8%
2.	Moderate halophiles: *Micrococcus halobios Pseudomonas, Bacillus, Planococcus halophilu Vibrio costicola, Desulfohalobium*	0.5 - 2.5 M 4.7 to 20%
3.	Borderline Extreme halophiles	1.5 – 4.0 M
4.	Extreme halophiles *Halobacterium, Halococcus, Salinibacter ruber*	2.5 – 5.2 M
5.	Halotolerant	2.5 M

5. **Halophiles:** Halophiles are organisms that thrive in high salt concentrations. The name comes from the Greek word for "salt-loving". Halophiles are classified into the Archaea domain, bacterial halophiles and eukaryota, such as the alga *Dunaliella salina* or fungus *Wallemia ichthyophaga*. Some well-known species give off a red color from carotenoid compounds, notably bacteriorhodopsin. Halophiles can be found anywhere with a concentration of salt five times greater than the salt concentration of the ocean, such as the Great Salt Lake in Utah, Owens Lake in California, the Dead Sea, and in evaporation ponds. Halophiles are categorized as slight, moderate, or extreme, by the extent of their halotolerance. Halophiles are isolated on salt water agar media which contain 3.0 - 4.0 % NaCl and other mineral salts.

   Properties of halophilic microorganisms and their negatively charged enzymes make them potentially very useful for biotechnology. Haloarchaea were the first members of archaea found to produce bacteriocins, named halocins. Carotene rich halobacteria and halophilic algae are used as food additives or as food-colouring agents, they improve dough quality of backing breed. The halophilic alga *Dunaliella salina* is used in the commercial source of β-carotene, potential source of glycerol production and as raw material for biofuel production. Halophilic bacteria produce Poly-β-hydroxy alkanoate which is used in the production of biodegradable plastics.

6. **Barophiles/ Piezophiles:** The term Barophil was coined by Zobell and Johnson in 1949. A piezophile (from Greek "piezo-" for pressure and "-phile" for loving) is an organism with optimal growth under high hydrostatic pressure. These organisms have maximum rate of growth at a hydrostatic pressure equal or above 10 MPa (= 99 atm = 1,450 psi). There are three types of barophiles:

   (i) **Barotolerants (facultative):** Grows at pressures from 100 – 500 atm.

   (ii) **Barophilic (obligative):** Grows at pressure from 400 – 500 atm.

   (iii) **Extreme Barophilic:** Grows at pressures higher than 500 atm.

Piezophilic bacteria have a high proportion of fatty acids in their cytoplasmic membrane, which allows membranes to remain functional and keep from gelling at high pressures. As the deep sea is also a cold environment, barophiles generally exhibit psychrophilic properties as well. Thermophilic barophilic or barotolerant bacteria are associated with deep-sea hot vents. *Thermococcus barophilus*, a hyperthermophile isolated from a Mid-Atlantic Ridge hydrothermal vent, requires high pressure when grown at the 100°C temperature. Other examples include *Shewanella benthica, Shewanella violacea, Photobacterium profundum, Moritella japonica, Moritella yayanosii. Pseudomonas bathycetes* grows at 1000 atm pressure and 3°C. Some barophiles have been found at the bottom of the Pacific Ocean (Mariana Trench-10500 m) where pressure often exceeds 117 MPa. These organisms cannot grow at pressure below 400-500 atm. Barophiles are also found in deep oil and sulphur wells at 400 psi pressure and 60 – 105°C temperature. For the growth of Barophiles, special vessels are designed which are pressurized to high degree. Only liquid media of capacity 4 – 5 ml are used in these vessels.

1.4.3 Cultivation of Chemolithotrophic Bacteria

Chemolithotrophic bacteria obtain their energy from the oxidation of inorganic (non-carbon) compounds. They derive their energy from the energy already stored in chemical compounds. By oxidizing the compounds, the energy stored in chemical bonds can be utilized in cellular processes. Examples of inorganic compounds that are used by these types of bacteria are sulfur, ammonium ion (NH^{4+}), and ferrous iron (Fe^{2+}). Chemolithotrophs can grow on medium that is free of carbon. The designation lithotrophic means "rock eating," the ability of these bacteria to grow in inhospitable environments.

1. **Colorless sulfur bacteria:** These bacteria oxidize hydrogen sulphide (H_2S) by accepting an electron from the compound. The acceptance of an electron by an oxygen atom creates water and sulphur. The energy from this reaction is then used to reduce carbon dioxide to create carbohydrates. An example of a colorless sulfur bacteria is the genus *Thiothrix*.

2. **Iron bacteria:** These bacteria are most commonly encountered as the rusty coloured and slimy layer that builds up on the inside of

toilet tanks. In a series of chemical reactions that is like those of the sulphur bacteria, iron bacteria oxidize iron compounds and use the energy gained from this reaction to drive the formation of carbohydrates. Examples of iron bacteria are *Thiobacillus ferrooxidans* and *Thiobacillus thiooxidans*. These bacteria are common in the runoff from coal mines. The water is very acidic and contains ferrous iron.

3. **Nitrifying bacteria:** They oxidize ammonia (NH_3) to nitrate (NO_3). Plants can use the nitrate as a nutrient source. These nitrifying bacteria are important in the operation of the global nitrogen cycle. Examples: *Nitrosomonas* and *Nitrobacter*.

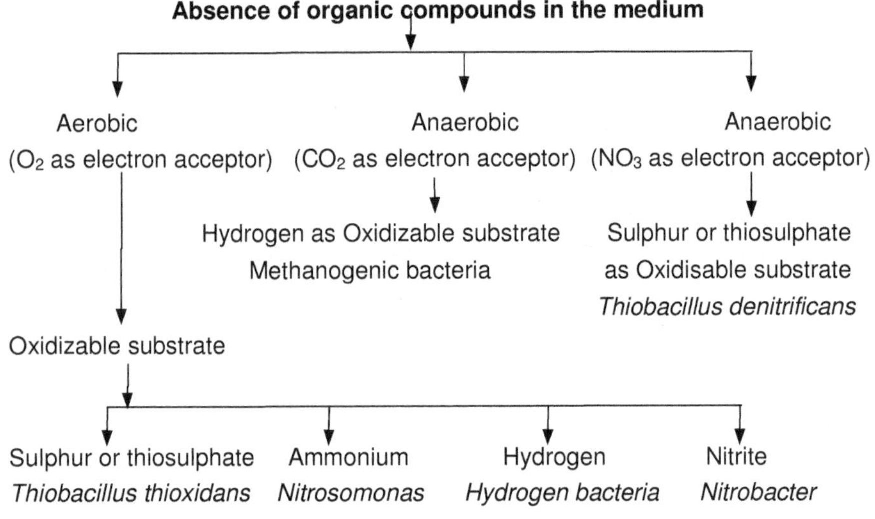

1. **Nitrifying bacteria:**

   i. Ammonia **oxidisers:** *Nitrosomonas, Nitrospira, Nitrosococcus, Nitrosolobus*

$$NH_4^+ \longrightarrow NO_2^-$$

   ii. **Nitrate oxidisers:** *Nitrobacter, Nitrococcus*

$$NO_2^- \longrightarrow NO_3^-$$

2. **Hydrogen bacteria:** *Hydrogenomonas*

$$H_2 \longrightarrow 2H^+ + 2e^-$$

Cultivation of *Thiobacillus ferroxidans*

9K Medium

Solution A

$(NH_4) 2SO_4$	3.0 gm
KCl	0.1 gm
K_2HPO_4	0.5 gm
$MgSO_4. 7H_2O$	0.5 gm
$Ca(NO_3)_2$	0.01 gm
Distilled water	700.0 ml

Solution B

$FeSO_4. 7H_2O$	44.22 gm
H_2SO_4 (1N)	10.0 ml
Distilled water	290.0 ml

Solutions A and B are sterilized separately and then mixed aseptically.

1.4.4 Cultivation of Anaerobes

Following four methods are used for Cultivation of Anaerobes:

1. Use of Special Anaerobic Culture Media
2. Anaerobic Chamber
3. Anaerobic Bags or Pouches
4. Anaerobic Jars.

1. Special Anaerobic Culture Media:

Of all the methods available for the cultivation of anaerobic bacteria, exclusion of oxygen from the medium is the simplest method. During preparation, the liquid culture medium is boiled by holding in a boiling water both for 10 minutes to drive off most of the dissolved oxygen. Liquid media soon become aerobic thus a reducing agent (e.g., cysteine 0.1%, ascorbic acid 0.1%, sodium thioglycollate 0.1%), is added to further lower the oxygen content. Oxygen-free N_2 is bubbled through the medium to maintain anaerobic condition. The medium is then dispensed into tubes, which are stoppered tightly and sterilized by autoclaving. Such tubes can be stored for many months before being used. During inoculation, the tubes are continuously flushed with oxygen free CO_2.

Special anaerobic culture medium for obligate anaerobic bacteria: Cooked meat broth (CMB; original medium known as 'Robertson's bullock-heart medium') and thioglycollate broth and its modifications are also very useful. CMB is suitable for growing anaerobic bacteria in air and for the preservation of their stock cultures.

2. Anaerobic Chamber:

Anaerobic chamber is an ideal anaerobic incubation system, which provides oxygen free environment for inoculating media and incubating cultures. It refers to a plastic anaerobic glove box that contains an atmosphere of H_2, CO_2, and N_2. Culture media are placed within the air-lock with the inner door. Air of the chamber is removed by a vacuum pump connection and replaced with N_2 through outer doors. The culture media are now transferred from air-lock to the main chamber, which contains an atmosphere of H_2, CO_2, and N_2. A circulator fitted in the main chamber circulates the gas atmosphere through pellets of palladium catalyst causing any residual O_2 present in the culture media to be used up by reaction with H_2. When the culture media become completely anaerobic they are inoculated with bacterial culture and placed in an incubator fitted within the chamber. The function of CO_2 present in the chamber is that it is required by many anaerobic bacteria for their best growth.

3. Anaerobic Bags or Pouches:

Anaerobic bags or pouches make convenient containers when only a few samples are to be incubated anaerobically. They are available commercially. Bags or pouches have an oxygen removal system consisting of a catalyst and calcium carbonate to produce an anaerobic, CO_2-rich atmosphere.

One or two inoculated plates are placed into the bag and the oxygen removal system is activated and the bag is sealed and incubated. Plates can be examined for growth without removing the plates from bag, thus without exposing the colonies to oxygen.

But as with anaerobic jar, plates must be removed from the bags in order to work with the colonies at the bench. These bags are also useful in transport of biopsy specimen for anaerobic cultures.

4. Anaerobic Jars (or GasPak Anaerobic System):

When an oxygen-free or anaerobic atmosphere is required for obtaining surface growth of anaerobic bacteria, anaerobic jars are the best suited. The most reliable and widely used anaerobic jar is the Melntosh-Fildes' anaerobic jar. It is a cylindrical vessel made of glass or metal with a metal lid, which is held firmly in place by a clamp.

Construction of Anaerobic Jar:

The lid possesses two tubes with taps, one acting as gas inlet and the other as the outlet. On its under surface it carries a gauze sachet carrying palladium pellets, which act as a room temperature catalyst for the conversion of hydrogen and oxygen into water. Palladium pellets act as catalyst, as long as the sachet is kept dry.

Inoculated culture plates are placed inside the jar and the lid clamped tight. The outlet tube is connected to a vacuum pump and the air inside is evacuated. The outlet tap is then closed and the gas inlet tube connected to a hydrogen supply. Hydrogen is drawn in rapidly. As soon as this inrush of hydrogen gas has ceased the inlet tube is also closed.

After about 5 minutes inlet tube is further opened. There occurs again an immediate inrush of hydrogen since the catalyst creates a reduced pressure within the jar due to the conversion of hydrogen and leftover oxygen into water.

If there is no inrush of hydrogen, it means the catalyst is inactive and must be replaced. The jar is left connected to the hydrogen supply for about 5 minutes, then the inlet tube is closed and the jar is placed in the incubator. Catalysis will continue until all the oxygen in the jar has been used up.

The GasPak is now the method of choice for preparing anaerobic jar. The GasPak is commercially available as a disposable envelope containing chemicals, which generate hydrogen and carbon dioxide when water is added. After the inoculated plates are kept in the jar, the GasPak envelope with water added, is placed inside and the lid screwed tight.

Hydrogen and carbon dioxide are liberated and the presence of a cold catalyst in the envelope permits the combination of hydrogen and oxygen to produce an anaerobic environment.

The outstanding feature of the GasPak system is the disposable gas generator envelope, which does away with the need for a vacuum pump and cylinders of compressed gas; the operation of the jar is consequently very quick and simple. As the standard GasPak jar is not evacuated before use a relatively large volume of water is formed during catalysis.

An indicator should be used for verifying the anaerobic condition in the jar and for this purpose methylene blue is generally used. When it is placed in an anaerobic environment, it is reduced from its coloured oxidized form to a colourless reduced leuco-compound.

Removal of the culture plates from the jar for microscopic examination is the major disadvantage of any anaerobic jar system. This, of course, results in the exposure of the colonics to oxygen, which is especially hazardous to the anaerobes during their first 48 hours of growth. For this reason, a suitable oxygen-free holding system always should be used in conjunction with anaerobic jars.

The culture plates should be removed from the jar and placed in the oxygen-free holding system. From there they should be removed one by one for rapid microscopic examination of colonies, and then quickly returned to the holding system. Plates never should remain in room air on the open bench.

1.4.5 Cultivation of Algae

Algae is a large, diverse group of photosynthetic eukaryotic organisms which includes organisms from unicellular microalgae, such as chlorella and the diatoms, to multicellular forms, such as the giant kelp, a large brown alga which may grow up to 50 m in length. Most are aquatic and lack many of the distinct cell and tissue types, such as stomata, xylem, and phloem, which are found in land plants. The largest and most complex marine algae are called seaweeds, while the most complex freshwater forms are the charophyta, a division of green algae which includes, for example, *Spirogyra*. Prokaryotic organisms cyanobacteria are also known as blue-green algae. Algae are capable of photosynthesis and produce their own nourishment by using light energy from the sun and carbon dioxide in order to generate carbohydrates and oxygen. most algae are photoautotrophs (reflecting their use of light energy to generate nutrients). However, there exist certain algal species that need to obtain their nutrition solely from outside sources; that is, they are heterotrophic. Such species acquire nutrients from organic materials (carbon containing compounds such as carbohydrates, proteins and fats). Algae are capable of reproducing through asexual or vegetative methods and via sexual reproduction. Asexual reproduction involves the production of a motile spore, while vegetative methods include simple cell division (mitosis) to produce identical offspring and the fragmentation of a colony. Sexual reproduction involves the union of gametes (produced individually in each parent through meiosis).

Cyanobacteria: These are also referred to as blue-green algae. They conduct oxygen-producing photosynthesis and live in many of the same environments as eukaryotic algae, cyanobacteria are gram-negative bacteria, and therefore are prokaryotes. They are also capable of independently conducting nitrogen fixation, the process of converting atmospheric nitrogen to usable forms of the element such as ammonia. The prefix "cyano" means blue. These bacteria have pigments that absorb specific wavelengths of light and give them their characteristic colours. Many cyanobacteria have the blue pigment phycocyanin, a light-harvesting pigment (it absorbs red wavelengths of light). Cyanobacteria all have some form of the green pigment chlorophyll, which is responsible for harvesting light energy during the photosynthetic process. Some others also have the red pigment phycoerythrin, which absorbs light with the green region and bestows the bacteria with a pink or red colour.

Cyanobacteria are cultivated by using BG -11 medium (per liter):

BG11 Broth is a universal medium for the cultivation and maintenance of blue green algae (cyanobacteria).

Sodium nitrate	1.5 gm
Dipotassium hydrogen phosphate	0.031 gm
Magnesium sulphate	0.036 gm
Calcium chloride dihydrate	0.0367gm
Sodium carbonate	0.020 gm
Disodium magnesium EDTA	0.001 gm
Citric acid	0.0056 gm
Ferric ammonium citrate	0.006 gm
Final pH after sterilization (at 25°C)	7.1

Autotrophic microalgae are cultivated on land in large ponds, or in enclosed so-called photobioreactors, using enriched CO_2.

Heterotrophic microalgae are grown in large fermenters using sugar or starch.

Seaweeds (macroalgae) are cultivated in seawater, typically in near-shore systems, though open ocean.

For the cultivation of algae growth requirements are light, CO_2, H_2O, nutrients (mainly Nitrogen : Phosphorus : Potassium) and trace elements. By means of photosynthesis the alga synthesizes all the biochemical compounds necessary for growth. Some are unable to synthesize certain biochemical compounds (e.g. Vitamins) and will require these to be present in the medium. This condition is known as auxotrophy.

Spirulina nutrient medium: Prepare solutions I and II; Autoclave solutions I and solution II with trace metals separately and cool; aseptically combine the both solutions. Aseptically add 1 ml of the cyanocobalamin (B_{12}) solution.

Solution I (500 ml):

$NaHCO_3$	13.61 gm
Na_2CO_3	4.03 gm
K_2HPO_4	0.50 gm

Solution II (500 ml):

$NaNO_3$	2.50 gm
K_2SO_4	1.0 gm
NaCl	1.0 gm
$MgSO_4$ 7(H_2O)	0.20 gm
$CaCl_2$ 2(H_2O)	0.04 gm
Trace Metals	1 ml ($FeSO_4$ 7(H_2O) 0.7 gm, $ZnSO_4$ 7(H_2O) 1.0 gm, $MnSO_4$ 7(H_2O) 2 gm, Co $(NO_3)_2$ 6(H_2O) 1 gm, $CuSO_4$ 5(H_2O) 0.005 gm, 1 lit Distilled water)
Cyanocobalomin	1 ml (0. 5 mg in 1 lit Distilled water)

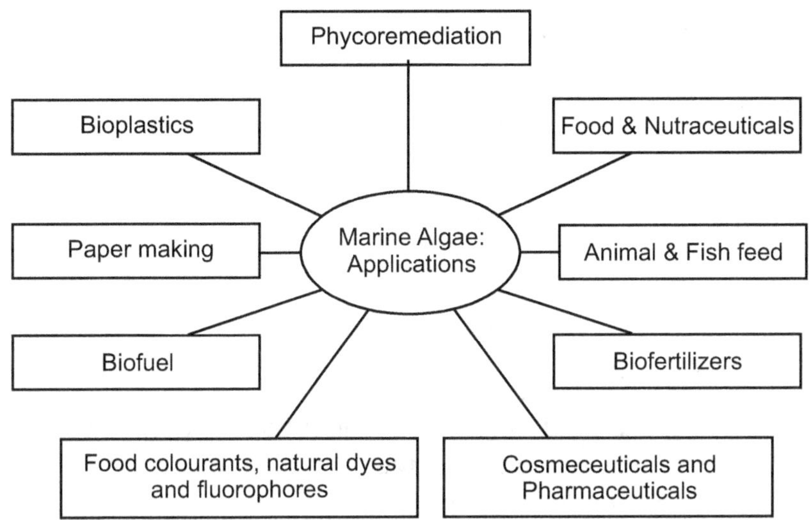

Fig. 1.8

1.4.6 Cultivation of Fungi

All fungi are heterotrophs. For their cultivation, simple media containing sugars, inorganic nitrogen, other inorganic salts supplying phosphates and other minerals can be used. Complex media containing organic ingredients like peptone are used. Media used for fungal culture are Sabouraud's dextrose, malt extract and brain heart infusion medium. To prevent contamination of the medium by bacteria, chloramphenicol is used, but prevents the growth of Actinomyces. In general, to cultivate yeasts and fungi high sugar content (3-4%) and low pH (3.5 -6) is used.

1. **Sabouraud's Agar:**

Glucose	40 g
Peptone	10 g
Agar	15 g
Distilled water	1000 ml

This is the standard medium used in medical mycology. It can be used as a general laboratory medium in place of Leonian's or PDA, although the amount of glucose is rather high and may suppress sporulation in some fungi.

2. Czapek-Dox Thom Agar:

Sucrose	30.0 g
$NaNO_3$	3.0 g
K_2HPO_4	1.0 g
$MgSO_4.7H_2O$	0.5 g
KCl	0.5 g
$FeSO_4.7H_2O$	0.01 g
Agar	15.0 g
Distilled water	1000 ml

Nitrate serves as the sole source of nitrogen. It is a synthetic medium particularly used for the identification of species of *Aspergillus* and *Penicillium*. Many moulds produce very characteristic colonies on it and may also exude pigmented substances. Aerial growth is suppressed, and sporulation is enhanced.

3. Potato Dextrose Agar:

Thinly sliced, peeled white potatoes	500 g
Glucose	20 g
Agar	15 g
Distilled water	1000 ml

Heat potatoes at 60°C for 1 hour and filter through cheese cloth. Make up volume to 1000 ml and add other ingredients. Cook 1 hour and then sterilize. Potato Dextrose Agar, or PDA, has been used from olden days by plant pathologists and many mycologists for general laboratory use.

4. Modified Leonian's Agar:

Maltose	6.25 g
Malt extract	6.25 g
KH_2PO_4	1.25 g
Yeast extract	1.0 g
$MgSO_4.7H_2O$	0.625 g
Peptone	0.625 g
Agar	20 g
Distilled water	1000 ml

Leonian's medium was devised by the American mycologist L.H. Leonian to promote sporulation in certain moulds. R.F. Cain from University of Toronto found it to be more suitable for ascomycetes if it contained a little yeast extract; hence the term "modified" in its name. Certain fungi, such as many mycorrhizal forms, do not grow on Leonian's Agar because they are unable to use maltose as an energy source. For these fungi modified Melin-Norkran's medium is used, which has a glucose energy component.

5. **Dextrose-peptone-yeast extract Agar (DPYA):**

Glucose (dextrose)	5.0 g
Peptone	1.0 g
Yeast extract	2.0 g
NH_4NO_3	1.0 g
K_2HPO_4	1.0 g
$MgSO_4.7H_2O$	0.5 g
$FeCl_3.6H_2O$	0.01 g
Oxgall	5.0 g
Sodium propionate	1.0 g
Chlortetracycline	30 mg
Streptomycin	30 mg
Agar	20 g
Distilled water	1000 ml

DPYA is an excellent medium for isolating fungi from soil, dung and other natural substrates. The oxgall and sodium propionate restrict the growth of some rapidly spreading fungi, while the chlortetracycline and streptomycin discourage bacteria. It can also be prepared as a more general-use medium by omitting these four inhibitory substances.

1.4.7 Cultivation of Actinomycetes

Actinomycetes are classified as a group of gram-positive bacteria that are unique for their spore forming abilities and formation of mycelia structures. Actinomycetes are filamentous bacteria that show fungi like growth. Actinomycetes can be isolated from soils and other natural substrates on suitable agar media. Actinomycete colonies can easily be distinguished on the plate from those of fungi and true bacteria. They are compact, often leathery giving a conical appearance, and have a dry surface. Actinomycetes have provided many industrially important bioactive compounds having great economic importance and always being a curious organism for secondary metabolite production. Actinomycetes have long been recognized as prolific producers of enzymes, antibiotics, anti-cancerous agents and play an important role in recycling of organic matter. Several antibiotics including erythromycin, streptomycin, rifamycin and gentamycin, are all products isolated from soil actinomycetes. Several culture media are used for isolation of actinomyces.

1. Starch Casein Agar (SCA):

Starch	10 gm
K_2HPO_4	2 gm
KNO_3	2 gm
Casein	0.3 gm
$MgSO_4.7H_2O$	0.05 gm
$CaCO_3$	0.02 gm
$FeSO_4.7H_2O$	0.01 g
Agar	15 gm
Distilled water	1000 ml
pH	7.0 ± 0.1

2. Luria Bertani Medium

Starch	10 gm
Peptone	2.0 gm
Yeast Extract	4.0 gm
Agar	18.0 gm
Distilled water	1000 ml
pH	7.0 ± 0.1

3. Starch Nitrate Agar

Soluble starch	20.0 g
K_2HPO_4	1.0 gm
KNO_3	2.0 gm
$MgSO_4$	0.5 gm

$CaCO_3$	3.0 gm
NaCl	10.0 gm
$FeSO_4O$	1 gmm
$MnCl_2$	0.1 g
$ZnSO_4$	0.1 gm
Distilled water	1000 ml
pH	7.0 ± 0.1

Other media includes Potassium tellurite agar medium, Half – strength nutrient agar medium, Oatmeal agar medium, Yeast extract-malt extract agar medium, Inorganic salt – starch agar medium, Glucose asparagine agar medium, Arginine-Glycerol medium, Tyrosine agar etc.

1.4.8 Cultivation of Viruses

Viruses are obligate intracellular parasites, so they depend on host for their survival. They cannot be grown in non-living culture media or on agar plates alone, they must require living cells to support their replication.

The primary purposes of virus cultivation are:

1. To isolate and identify viruses in clinical samples.
2. To do research on viral structure, replication, genetics and effects on host cell.
3. To prepare viruses for vaccine production.

Cultivation of viruses is carried out by using:

1. Animal Inoculation
2. Inoculation into embryonated egg
3. Cell Culture

1. Animal Inoculation:

- Viruses which are not cultivated in embryonated egg and tissue culture are cultivated in laboratory animals such as mice, guinea pig, hamster, rabbits and primates.
- The selected animals should be healthy and free from any of communicable diseases.
- Inoculation can be done by intracerebral, intranasal, intraperitoneal and subcutaneous routes.

- After inoculation, virus multiply in host and develops disease. The animals are observed for symptoms of disease and death.
- Then the virus is isolated and purified from the tissue of these animals.

Advantages of Animal Inoculation:

- Diagnosis, Pathogenesis and clinical symptoms are determined.
- Production of antibodies can be identified.
- Primary isolation of certain viruses.
- Mice provide a reliable model for studying viral replication.
- Used for the study of immune responses, epidemiology and oncogenesis.

Disadvantages of Animal Inoculation:

- Expensive and difficulties in maintenance of animals.
- Difficulty in choosing of animals for virus.
- Some human viruses cannot be grown in animals or can be grown but do not cause disease.
- Issues related to animal welfare systems.

2. **Inoculation into Embryonated Egg:**

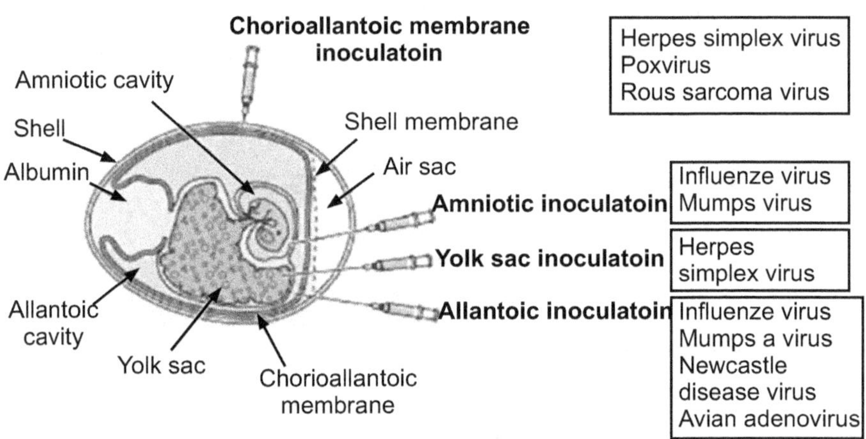

Fig. 1.9

- Good Pasture in 1931 first used the embryonated hen's egg for the cultivation of virus.
- The process of cultivation of viruses in embryonated eggs depends on the type of egg which is used.

- Viruses are inoculated into chick embryo of 7-12 days old.
- For inoculation, eggs are first prepared for cultivation, the shell surface is first disinfected with iodine and penetrated with a small sterile drill.
- After inoculation, the opening is sealed with gelatine or paraffin and incubated at 36°C for 2-3 days.
- After incubation, the egg is broken, and virus is isolated from tissue of egg.
- Viral growth and multiplication in the egg embryo is indicated by the death of the embryo, by embryo cell damage, or by the formation of typical pocks or lesions on the egg membranes.
- Viruses can be cultivated in various parts of egg like chorioallantoic membrane (for pox virus, Herpes virus), allantoic cavity (influenza virus, yellow fever, rabies), amniotic sac (influenza virus and the mumps virus) and yolk sac (`1for some viruses and some bacteria like Chlamydia, Rickettsiae).

Advantages of Inoculation into Embryonated Egg:

- Widely used method for the isolation of virus and growth.
- Ideal substrate for the viral growth and replication.
- Isolation and cultivation of many avian and few mammalian viruses.
- Cost effective and maintenance is much easier.
- Less labour is required.
- The embryonated eggs are readily available.
- They are free from contaminating bacteria and many latent viruses.
- Widely used method to grow virus for some vaccine production.

Disadvantage of Inoculation into Embryonated Egg:

- The site of inoculation varies with different virus because each virus has different site for its growth and replication.

3. Cell Culture (Tissue Culture):

There are three types of tissue culture; organ culture, explant culture and cell culture.

Organ cultures are mainly done for highly specialized parasites of certain organs e.g. tracheal ring culture is done for isolation of coronavirus.

Explant culture is rarely done.

Cell culture is mostly used for identification and cultivation of viruses.

- Cell culture is the process by which cells are grown under controlled conditions.

- Cells are grown *in-vitro* on glass or a treated plastic surface in a suitable growth medium.

- At first growth medium, usually balanced salt solution containing 13 amino acids, sugar, proteins, salts, calf serum, buffer, antibiotics and phenol red are taken and the host tissue or cell is inoculated.

- On incubation the cell divides and spread out on the glass surface to form a confluent monolayer.

Types of Cell Culture:

1. Primary Cell Culture:

- These are normal cells derived from animal or human cells.

- They can grow only for limited time and cannot be maintained in serial culture.

- They are used for the primary isolation of viruses and production of vaccine.

- Examples: Monkey kidney cell culture, Human amnion cell culture.

2. Diploid Cell Culture (Semi-continuous cell lines):

- They are diploid and contain the same number of chromosomes as the parent cells.

- They can be sub-cultured up to 50 times by serial transfer following senescence and the cell strain is lost.

- They are used for the isolation of some fastidious viruses and production of viral vaccines.

- Examples: Human embryonic lung strain, Rhesus embryo cell strain.

3. Heteroploid Cultures (Continuous cell lines):

- They are derived from cancer cells.

- They can be serially cultured indefinitely so named as continuous cell lines.

- They can be maintained either by serial subculture or by storing in deep freeze at −70°C.
- Due to derivation from cancer cells they are not useful for vaccine production.
- Examples: HeLa (Human Carcinoma of cervix cell line), HEP-2 (Humman Epithelioma of larynx cell line), Vero (Vervet monkey) kidney cell lines, BHK-21 (Baby Hamster Kidney cell line).

Advantage of Cell Culture:

- Relative ease, broad spectrum, cheaper and sensitive.

Disadvantage of Cell Culture:

- The process requires trained technicians with experience.

Cultivation of Plant Viruses:

There are some methods of cultivation of plant viruses such as plant tissue cultures, cultures of separated cells, or cultures of protoplasts, etc. viruses can be grown in whole plants. Leaves are mechanically inoculated by rubbing with a mixture of viruses and an abrasive. When the cell wall is broken by the abrasive, the viruses directly contact the plasma membrane and infect the exposed host cells. A localized necrotic lesion often develops due to the rapid death of cells in the infected area. Some plant viruses can be transmitted only if a diseased part is grafted onto a healthy plant.

Cultivation of Bacteriophages (Viruses that infects bacteria):

Bacteriophages were discovered more than a century ago. In 1896, Ernest Hanbury Hankin, a British bacteriologist reported that something in the waters of rivers in India had unexpected antibacterial properties against cholera and this water could pass through a very fine porcelain filter and keep this distinctive feature.

Bacteriophages (phage) are obligate intracellular parasites. They multiply inside a bacterium by making use of some or all of the host (i.e., bacteria) biosynthetic machinery. They enter the bacterial cell by 'landing' on the cell wall and injecting their DNA into the bacterial cytoplasm. After entry, the phage DNA acts as a template for production of phage proteins. These proteins replicate the phage and subjugate the cell, eventually causing lysis and death of the host cell. A bacteriophage particle is even harder to see than a bacterium. Viruses are beyond the limits of resolution of the light microscope and can be seen only with electron microscopes. Fortunately, we can use a technique very similar to the colony-counting technique used to measure the number of bacteria to count phage particles, known as the plaque assay. Lytic phages are enumerated by this method.

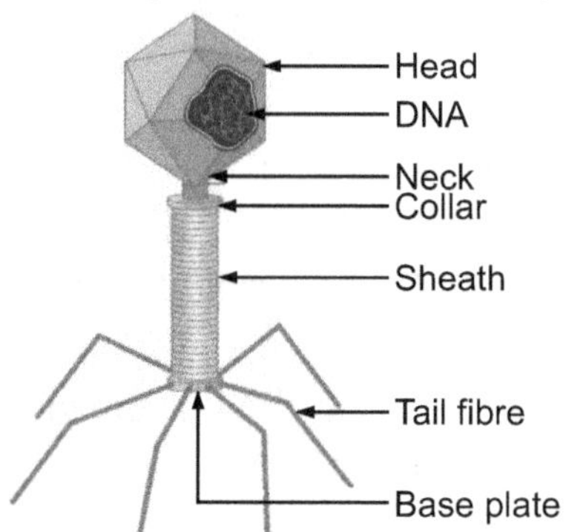

Fig. 1.10: A typical phage

The plaque assay is employed to count and measure the infectivity level of the bacteriophages. This assay is the most widely used technique for the isolation of virus and its purification, and to optimize the viral titers. The basis of plaque assay is to measure the ability of a single infectious virus to form a "plaque" on a concurrent monolayer culture cells. A plaque is developed as a part of infection of one cell by a single virus particle that is followed by the replication of that virus, and finally, the death of the cell. The newly replicated virus particles will later infect and then kill surrounding cells.

For this technique, virulent phage stock and a susceptible host cell culture is required. 10-fold dilutions of the phage stock are prepared. The procedure requires the use of a Double-Layer Agar (DLA) technique also known as double agar overlay method, in which the hard agar serves as a base layer (to form gel), and a mixture of few phage particles (diluted stock) and a very large number of host cells in a soft agar forms the upper overlay. When the plates are incubated, susceptible *Escherichia coli* cells multiply rapidly and produce a lawn of confluent growth on the medium. When one phage particle adsorbs to a susceptible cell, penetrates the cell, replicates and release new phage particles which infect other bacteria in the vicinity of the initial host cell. The growth or spread of the new viruses is then restricted or limited to the neighbouring cells by the gel. This cycle is repeated until large numbers of bacteria have been destroyed. The destroyed cells produce single circular, non-turbid areas called plaques in the bacterial lawn, where there is no growth of bacteria because the phage progeny originating from single virus particles have multiplied sufficiently to kill bacteria over an easily visible area. Eventually the plaque becomes too large to be visible to our naked eye. Each plaque represents the lysis of a phage-infected bacterial culture and can be designated as a plaque-forming unit (PFU) and is used to quantitate the number of infective phage particles in the culture. Dyes that stain the living cells are frequently used to enhance the contrast between the plaques and the living cells. Therefore, the dead cells in the plaque will appear as unstained against the colored background. Only viruses that can cause visible damage of cells can be assayed using this way.

The number of phage particles contained in the original stock phage culture is determined by counting the number of plaques formed on the seeded agar plate and multiplying this by the dilution factor.

All known phages can be divided in two groups according to the type of infection. One group is characterised by a lytic infection and the other is represented by a lysogenic, or temperate, type of infection.

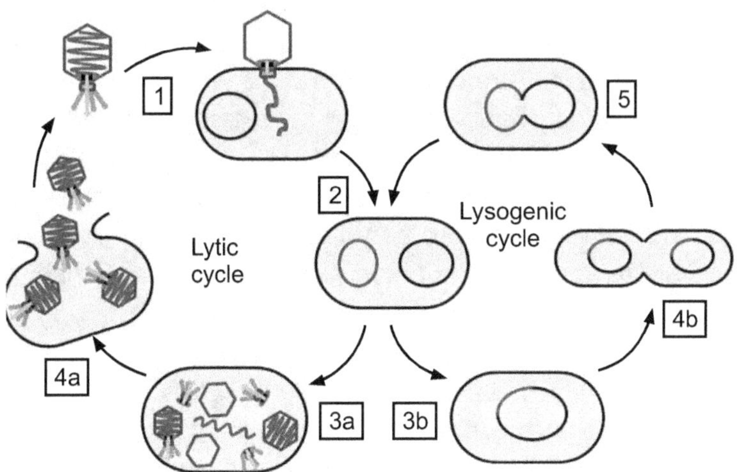

Fig. 1.11: Two cycles of bacteriophage reproduction

1 - Phage attaches the host cell and injects DNA; 2 - Phage DNA enters lytic or lysogenic cycle; 3a – New phage DNA and proteins are synthesised and virions are assembled; 4a – Cell lyses releasing virions; 3b and 4b – steps of lysogenic cycle: integration of the phage genome within the bacterial chromosome (becomes prophage) with normal bacterial reproduction; 5 - Under certain conditions the prophage excises from the bacterial chromosome and initiates the lytic cycle.

1.5 CONCEPT OF PURE CULTURE, ENRICHMENT, ISOLATION

Pure culture consists of a nutrient medium containing the growth of a single species of organism. A mixed culture consists of a nutrient medium containing the growth of two or more species of organisms. Pure cultures are required for studying the morphology and physiology of organisms. A pure culture is developed from a mixed culture (one containing many species) by transferring a small sample into new, sterile growth medium in such a manner as to disperse the individual cells across the medium surface. This method separates the individual cells so that, when they multiply, each will form a discrete colony, which is inoculated on another medium, to get the pure culture. A plate culture consists of organism growing on a solid medium contained in the petri dish. A slant culture consists of an organism growing on the inclined surface of the solid medium such as nutrient agar which is referred as nutrient agar slant culture. A solid medium in a slanted position increases the surface area and gives the heavier growth of organisms as compared to liquid media. A stab culture is prepared by stabbing a solid medium such as nutrient agar with a straight wire loop upto complete depth of the agar column. A liquid culture consists of a liquid medium such as nutrient broth containing the growth of organisms.

Enrichment Culture Method is used to isolate those microorganisms, which are present in relatively less numbers or that have slow growth rates compared to the other species present in the mixed culture. The enrichment culture strategy provides a specially designed cultural environment by incorporating a specific nutrient in the medium and by modifying the physical conditions like salt concentration, pH, temperature of the incubation etc. The medium of known composition and, specific condition of incubation favours the growth of desired microorganisms but, is unsuitable for the growth of other types of microorganisms. It is based on the principle of natural selection at microscale level to isolate bacteria which are unusual and less in number. E.g. To isolate thermophile, increase in the temperature of incubation will allow growth of thermophile only while others get killed at high temperature.

Pure cultures can be obtained using streak plate, spread plate or pour plate method.

1.5.1 Streak Plate Method

This method is used most commonly to isolate pure cultures of bacteria. A small amount of mixed culture is placed on the tip of an inoculation transfer loop/needle and is streaked across the surface of the agar medium. The successive streaks "thin out" the inoculum sufficiently and the micro-organisms are separated from each other. It is usually desirable to streak out a second plate by the same loop/needle without reinoculation. These plates are incubated to

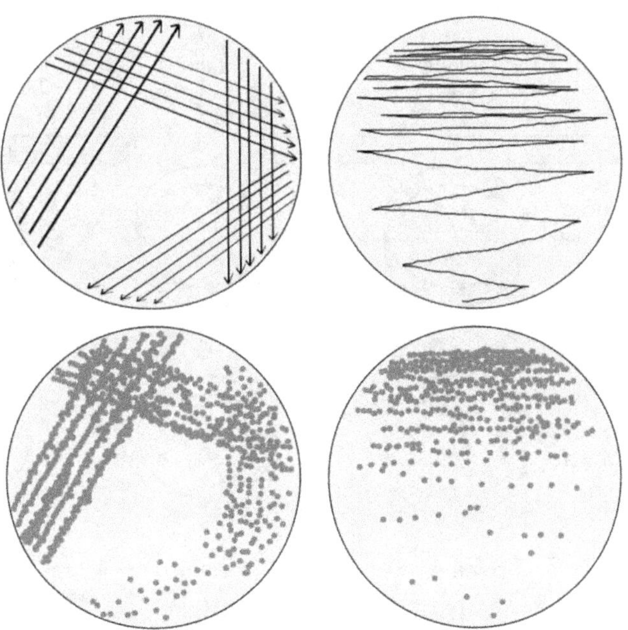

Fig. 1.12

allow the growth of colonies. The key principle of this method is that, by streaking, a dilution gradient is established across the face of the Petri plate as bacterial cells are deposited on the agar surface. Because of this dilution gradient, the bacterial cells remain sufficiently apart and the colonies developing from one cell will not merge with the colonies developing from another cell. Presumably, each colony is the progeny of a single microbial cell thus representing a clone of pure culture. Such isolated colonies are picked up separately using sterile inoculating loop/needle and re-streaked onto fresh media to ensure purity.

1.5.2 Pour Plate Method

This method involves plating of diluted samples mixed with melted agar medium. The main principle is to dilute the inoculum in successive tubes containing liquefied agar medium to permit a

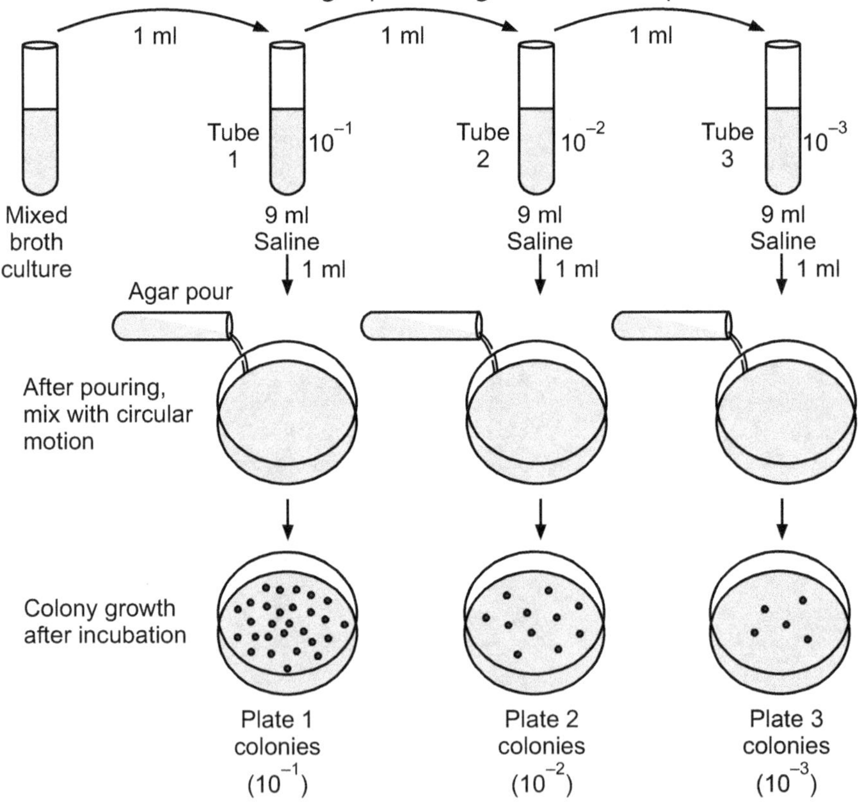

Fig. 1.13

thorough distribution of bacterial cells within the medium. In this method, the mixed culture of bacteria is diluted directly in tubes containing melted agar medium maintained in the liquid state at a temperature of 42-45°C (agar solidifies below 42°C). The bacteria and the melted medium are mixed well. The contents of each tube are poured into separate Petri plates, allowed to solidify, and then incubated. When bacterial colonies develop, one finds that isolated colonies develop both within the agar medium (subsurface colonies) and on the medium (surface colonies). These isolated colonies are then picked up by inoculation loop and streaked onto another Petri plate to insure purity.

Pour plate method has certain disadvantages as follows:

- The picking up of subsurface colonies needs digging them out of the agar medium thus interfering with other colonies, and

- The microbes being isolated must be able to withstand temporary exposure to the 42-45°C temperature of the liquid agar medium; therefore, this technique proves unsuitable for the isolation of psychrophilic microorganisms.

However, the pour plate method, in addition to its use in isolating pure cultures, is also used for determining the number of viable bacterial cells present in a culture.

1.5.3 Spread Plate Method

In this method, the mixed culture or microorganisms is not diluted in the melted agar medium (unlike the pour plate method); it is rather diluted in a series of tubes containing sterile liquid, usually, water or physiological saline (0.85%).

A drop of so diluted liquid from each tube is placed on the centre of an agar plate and spread evenly over the surface by means of a sterilized bent-glass-rod. The medium is now incubated. When the colonies develop on the agar medium plates, it is found that there are some plates in which well-isolated colonies grow. This happens as a result of separation of individual microorganisms by spreading over the drop of diluted liquid on the medium of the plate. The isolated colonies are picked up and transferred onto fresh medium to ensure purity. In contrast to pour plate method, only surface colonies develop in this method and the microorganisms are not required to withstand the temperature of the melted agar medium.

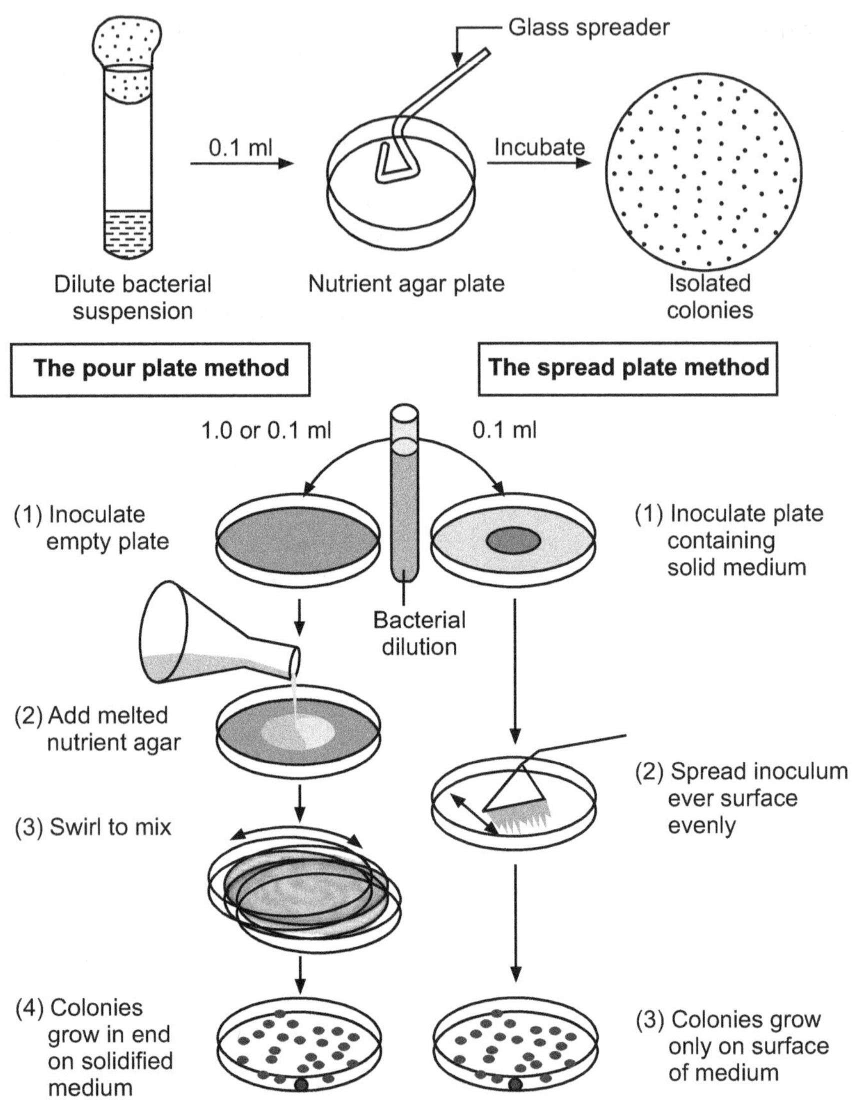

Fig. 1.14

1.5.4 Serial Dilution Method

This method is commonly used to obtain pure cultures of those microorganisms that have not yet been successfully cultivated on solid media and grow only in liquid media. A microorganism that predominates in a mixed culture can be isolated in pure form by a series of dilutions. The inoculum is subjected to serial dilution in a sterile liquid medium, and many tubes of sterile liquid medium are inoculated with aliquots of each successive dilution.

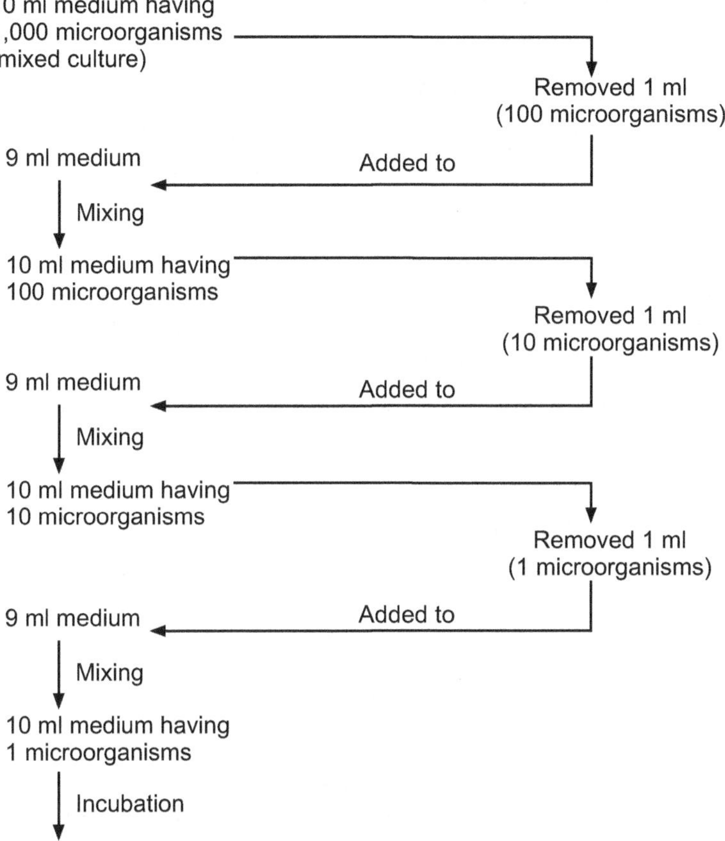

Fig. 1.15

The aim of this dilution is to inoculate a series of tubes with a microbial suspension so dilute that there are some tubes showing growth of only one individual microbe. For example, we have a culture containing 10 ml of liquid medium, containing 1,000 microorganisms i.e., 100 microorganisms/ml of the liquid medium. If we take out 1 ml of this medium and mix it with 9 ml of fresh sterile liquid medium, we would then have 100 microorganisms in 10 ml or 10 microorganisms/ml. If we add 1 ml of this suspension to another 9 ml of fresh sterile liquid medium, each ml would now contain a single microorganism. If this tube shows any microbial growth, there is a very high probability that this growth has resulted from the introduction of a single microorganism in the medium and represents the pure culture of that microorganism.

1.6 MAINTENANCE OF BACTERIAL AND FUNGAL CULTURES USING DIFFERENT TECHNIQUES

Once a microorganism has been isolated in pure culture it is important to preserve it in a viable state for further study and use. Microbiology laboratories maintain a large collection of strains which is referred as stock culture collection. Stock cultures must be maintained such that there is no loss of their biological, immunological and cultural characters.

The methods used to maintain the pure cultures include:

1. Periodic Transfer to Fresh Media
2. Refrigeration
3. Paraffin Method
4. Soil/grain cultures
5. Cryopreservation
6. Lyophilization.

1. Periodic Transfer to Fresh Media:

Strains can be maintained by periodically preparing a fresh culture from the previous stock culture. The culture medium, the storage temperature, and the time interval at which the transfers are made vary with the species and must be ascertained beforehand. The temperature and the type of medium chosen should support a slow rather than a rapid rate of growth so that the time interval between transfers can be if possible. Many of the more common heterotrophs remain viable for several weeks or months on a medium like **Nutrient Agar**. The transfer method has the disadvantage of failing to prevent changes in the characteristics of a strain due to the development of variants and mutants.

2. Refrigeration for Agar Slant Cultures:

Pure cultures can be successfully stored at 0-4°C either in refrigerators or in cold-rooms. This method is applied for short duration (2-3 weeks for bacteria and 3-4 months for fungi) because the metabolic activities of the microorganisms are greatly slowed down but not stopped. The pure culture suspension is subcultured on nutrient agar slope or slant in screw cap test tubes. Thus, their growth continues slowly, nutrients are utilized, and waste products released in medium. This results in, finally, the death of the microbes after some period.

3. **Paraffin Method / Preservation by Overlaying Cultures with Mineral Oil:**

This is a simple and most economical method of maintaining pure cultures of bacteria and fungi. In this method, sterile liquid paraffin oil is poured over the slant (slope) of culture so that the slant is completely immersed in the oil with a small layer at the top and stored upright at room temperature. The layer of paraffin ensures anaerobic conditions and prevents dehydration of the medium. This condition helps microorganisms or pure culture to remain in a dormant state and, therefore, the culture is preserved for several years. Such a method is used for unstable cultures which can lose their desired properties on frequent subculturing.

The advantage of this method is that we can remove some of the growth under the oil with a transfer needle, inoculate a fresh medium, and still preserve the original culture. The simplicity of the method makes it attractive, but changes in the characteristics of a strain can still occur.

4. **Soil/Grain Cultures:**

This method is used for aerobic spore forming bacterial and fungal cultures. In this method, soil is powdered, dried and sterilized. This soil is filled in the test tubes and resterilized. A small volume of dense spore suspension is added aseptically in the soil. These cultures are stored at room temperature without any loss in the viability for 2 – 3 years. A similar method is followed for grain cultures, where wheat, barley or corn can be used instead of soil.

5. **Cryopreservation:**

Cryopreservation (i.e., freezing in liquid nitrogen at –196°C) helps survival of pure cultures for long storage times i.e. for many years. In this method, the microorganisms of culture are rapidly frozen in liquid nitrogen at –196°C in the presence of stabilizing agents such as glycerol or Dimethyl Sulfoxide (DMSO) that prevent the cell damage due to formation of ice crystals and promote cell survival. This liquid nitrogen method has been successful with many species that cannot be preserved by lyophilization and most species can remain viable under these conditions for 10 to 30 years without undergoing change in their characteristics, however this method is expensive.

6. Lyophilization (Freeze-Drying):

In this method, vials containing thick suspension of culture are plugged with cotton wool and the culture is rapidly frozen at a very low temperature (-70°C to -120°C) and then dehydrated by vacuum. The frozen vials are put in the glass ampules. The glass ampules are attached to a vacuum pump and air and moisture is withdrawn from the vial and ampule. The top of the ampule is sealed. Under these conditions, the microbial cells are dehydrated, and their metabolic activities are stopped; as a result, the microbes go into dormant state and retain viability for years. Lyophilized or freeze-dried pure cultures are stored in the dark at 4°C in refrigerators. Freeze-drying method is used by all culture collection centres.

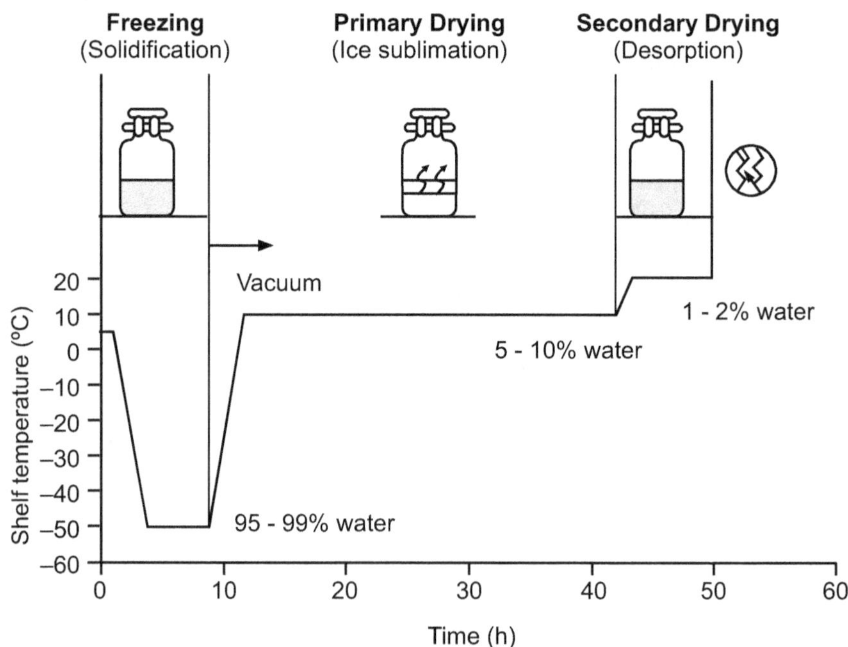

Fig. 1.16

Advantages of Lyophilization:
* Only minimal storage space is required; hundreds of lyophilized cultures can be stored in a small area.
* Small vials can be sent conveniently through the mail to other microbiology laboratories when packaged in a special sealed mailing container.
* Lyophilized cultures can be revived by opening the vials, adding liquid medium, and transferring the rehydrated culture to a suitable growth medium.
* Thermolabile materials can be dried.
* Sterility can be maintained.

Disadvantages of Lyophilization:
* Many biological molecules are damaged by the stress associated with freezing, freeze-drying, or both.
* The product is prone to oxidation, due to high porosity and large surface area. Therefore, the product should be packed in vacuum or using inert gas or in a container impervious to gases.
* Cost may be an issue, depending on the product.
* Long time-consuming process.

1.7 CULTURE COLLECTION CENTRES AND THEIR ROLE

Since, the early days of microbiology, enormous numbers of microorganisms were isolated from a wide variety of natural sources and used for the scientific research and for the industrial fermentation. However, large numbers of microorganisms had lost in the past, and they are no longer available. Microbiologists often lose microbial cultures that they studied because of the change of their interests and difficulties in keeping the cultures. Effective research needs adequate and reliable sources of properly preserved cultures. In the near future, uncountable numbers of microbial strains will be isolated through the study of microbial diversity, and the attributes of a large number of strains will be improved. Therefore, reliable and well-organized culture collections are needed as the depository and for the promotion of research and application of the strains. Social needs for culture collections are increasing year by year, and effective and smooth management is required for better services of culture collections.

Culture collections have more than a century old history. Prof. Frantisek Karl of Prague has been regarded as pioneer in this. He was the first one to collect the cultures and made them available to others. After his death in 1911, his collection was transferred to University of Vienna in 1915. Initially, the emphasis of cultures was for taxonomy and epidemiology. With advances in microbiology and biotechnology, modern day culture collections have much more diverse role. Their number has also increased significantly.The World Data Centre for Microorganisms (WDCM) was established 50 years ago at the data centre of the World Federation for Culture Collections (WFCC)—Microbial Resource Centre (MIRCEN). WDCM aims to provide integrated information services using big data technology for microbial resource centres and microbiologists all over the world. There are 647 culture collections in 70 countries registered with WDCM (World Data Centre for Microorganisms) database. Good infrastructure and expertise in long term preservation, characterization and identification of diverse group of microorganisms are necessary for a good service culture collection.

The World Federation for Culture Collections (WFCC) is a Multidisciplinary Commission of the International Union of Biological Sciences (IUBS) and a Federation within the International Union of Microbiological Societies (IUMS). The WFCC is concerned with the collection, authentication, maintenance and distribution of cultures of microorganisms and cultured cells. Its aim is to promote and support the establishment of culture collections and related services, to provide connection and set up an information network between the collections and their users, to organise workshops and conferences, publications and newsletters and work to ensure the long term perpetuation of important collections.

These culture collections are supported by different Government and private funding agencies. Among 647 culture collections 253 are supported by universities while 279 are supported by Government while 56, 39 and 20 are supported by semi-governmental organisations, private and industries. There are around 11 culture collections registered with WDCM from Africa, 153 from America, 220 from Asia, 22 from Oceania and 220 from Europe. Bacterial cultures account for 343,253 (42%), filamentous fungi for 372,304 (46%), viruses for 14,376 (1.8%), cell lines for 5,156 (0.6%), and others for 80,485 (9.9%) by the data of WDCM.

List of some culture collection centres:

1. National Collection of Industrial Bacteria, Aberdeen, Scotland, U.K. (NCIB).
2. National Collection of Type Cultures, London, U.K. (NCTC).
3. National Research Council, Ottawa, Canada. (NRC).
4. Northern Utilization Research and Development Division, U.S. Department of Agriculture, Illinois, U.S.A. (NRRL).
5. National Institute of Allergy and Infectious Diseases, Hamilton, Montana, U.S.A. (NIAID).
6. National Collection of Industrial Microorganisms, National Chemical Laboratory, Pune, India. (NCIM).
7. National Institute of Health, Japan. (NIHJ).
8. Institute of fermentation, Osaka, Japan. (IFO).
9. International Collection of Phytopathogenic Bacteria, Plant pathology laboratory, Harpendum, England, U.K.(NCPPB).
10. Department of Culture Collection, Institute of Biochemistry and Physiology of microorganisms, U.S. S. R. Academy of Sciences, Pushchino, Moscow region, U.S.S.R. (VKM).
11. Anaerobe Laboratory, Virginia Polytechnic Institute and State university, Blacksburg, Virginia, U.S.A. (VPI).
12. European Culture Collections' Organisation (ECCO), Denmark.
13. German Resource Centre for Biological Material (DSMZ), Germany.
14. Culture Collection, University of Goteborg (CCUG), Sweden.
15. American Type Culture Collection (ATCC), United States.
16. Japan Collection of Microorganisms (JCM), Japan.
17. World Federation for Culture Collections (WFCC).
18. National Centre for Microbial Resource, (NCMR) in Pune, India.

Role of culture collection centres:

Microbes constitute the largest biomass on the earth and comprise of three domains of life (bacteria, archaea and eukaryotes).

Almost 90% of this diversity is still unexplored. These microbes play an integral and unique role in the functioning of the ecosystems in maintaining a sustainable biosphere and productivity. They are responsible for nutrient recycling and detoxification. They act as biological control agents, biocatalysts and produce a wide variety of products that have pharmaceutical and industrial applications. They are also thought to be solutions for the food and energy crisis that the world may face in future. They play a fundamental role in improving the quality of life, alleviate poverty, malnutrition, etc. Advent of metagenomics era and projects like metagenomic analysis of Human Intestinal Tract, Human Microbiome Project and soil metagenome (Terragenome) have further emphasized their vital role on the planet and on human health. However, events like global warming, lifestyle changes and anthropogenic activities are resulting in the loss of microbial diversity. Culture collections play a crucial role in the conservation and sustainable use of microbial resources.

- The culture collection centres focus on basic research in microbial diversity, microbial taxonomy, microbial genomics and proteomics etc.

- Isolation and identification of microorganisms from various environmental niches.

- Preservation of microbial biodiversity from niche areas as metagenomic libraries.

- Development of new strategies for isolation of "not yet cultured" microbes.

- To provide consultation services for patent deposits, preservation, propagation, biodeterioration, industrial problems, biosystematics and microbial biodiversity issues etc.

- To serve as national facility for microbial culture collection, deposition, maintenance of important and useful microbial cultures and supply of the cultures on demand, and preparation of their informative documents.

- To serve as an International Depositary Authority (IDA).

- To establish and conduct workshops, seminars, symposia and training programmes in the area of microbial identification, preservation and advanced areas of microbial taxonomy and phylogeny.

- To provide the authentic biological material for high quality research and teaching in the form of reference strains, reagents for quality control, etc.
- Improvement of culture collections centre is critical and crucial for the further development of microbiology, microbial industry, and biotechnology.

1.8 REQUIREMENTS AND GUIDELINES OF NATIONAL BIODIVERSITY AUTHORITY FOR CULTURE COLLECTION CENTRES

The National Biodiversity Authority (NBA) was established by the Central Government in 2003 to implement India's Biological Diversity Act (2002). The NBA is a Statutory Body and it performs facilitative, regulatory and advisory functions for the Government of India on issues of conservation, sustainable use of biological resources and fair and equitable sharing of benefits arising out of the use of biological resources.

1. **General Authority for all the Specified culture collection centres:**

To act as Designated Repository (DR) for safe deposit of holotypes/isotypes/ paratypes of new taxa discovered in India and samples of biological resources accessed by foreign citizens/entities for research or sent/carried abroad by Indian citizens/institutions for research [vide Sections 39 (1-3) and 19 (1) of the Biological Diversity Act, 2002 read with Rule 14(6) (viii) of the Biological Diversity Rules, 2004, and clause 4(6) of the Regulation on ABS Guidelines notified in November 2014].

2. **Specific Mandates for the Designated Repositories:**

Sr. No.	Name of the Institution for Deposition of the microbes	Category
1.	National Institute of Oceanography, Goa	Marine flora and fauna
2.	National Bureau of Agriculturally Important Microorganisms, Mau Nath Bhanjan, U.P.	Agriculturally important microorganisms

Contd...

Sr. No.	Name of the Institution for Deposition of the microbes	Category
3.	Institute of Microbial Technology, Chandigarh.	Microorganisms for industrial use (actinobacteria, bacteria, fungi and yeasts)
4.	National Institute of Virology, Pune	Viruses
5.	Indian Agricultural Research Institute, New Delhi	Microbes/Fungi/Blue Green Algae
6.	Microbial Culture Collection, National Centre for Cell Science, Pune	Microbes (bacteria and fungi including yeasts, cell cultures)
7.	Botanical Survey of India, Kolkata.	Macrofungi, Macroalgae
8.	National Botanical Research Institute, Lucknow.	Macrofungi, Macroalgae
9.	Indian Council of Forestry Research and Education, Dehradun (Forest Research Institute, Dehradun; Institute of Forest Genetics and Tree Breeding, Coimbatore; and Tropical Forest Research Institute, Jabalpur).	Macrofungi, Macroalgae

3. Focal Points for the Designated Repositories:

The Head of the facility, where biological resources are deposited and conserved, shall be the Scientist. In charge of the Designated Repository and shall serve as the Focal Point for the NBA. Each DR is required to designate the Focal Point for this purpose.

4. Allocation of Unique Accession Number to the Deposits:

It will be done by the concerned Head of the Repository who shall also decide after the 5 year custody period whether the sample will be added to the national collection for long-term storage or discarded if the sample had been of a common occurrence as procured from the local market or appeared to be a duplicate of existing accessions. Wherever required, the NBA will intimate the repository concerned to extend the period beyond five years, in specific cases. Wherever any new taxon is deposited, it should be preserved for indefinite period.

5. Who are required to deposit types/specimens/samples of Biological Resources in DRs?

(i) Any person, who discovers a new taxon of biological resources occurring in India, is required to notify it to the relevant DR and deposit its holotype/isotype/paratype there [Section 39 (3) of the BD Act].

(ii) Persons/entities, as identified u/s 3(2), accessing biological resources occurring in India [Sub-rule 6 (viii) of Rule 14 of the Biological Diversity Rules].

(iii) Indian citizens/entities taking/sending abroad biological resource for non-commercial research by applying in Form B (Sub-regulation 1 of Regulation 13 of the ABS Guidelines notified in November 2014).

6. Where the Voucher Specimen/Reference Sample may be deposited?

Based on the type of biological resource, the voucher specimen/reference sample may be deposited in DRs as notified by Ministry of Environment, and Forests vide notifications F.No. 26-14/2007-CSC, dated August 28, 2008; F.No. 26-15/2007-CSC, dated September 12, 2012; F. No. 26-15/2007-CSC, dated July 8, 2013, as mentioned in table of clause No. 2 of these guidelines.

7. Ownership of the Deposits and Duration of Safekeeping:

Under the Biological Diversity Act 2002, Government of India has sovereign rights over the biological resources occurring in or obtained from India. Accordingly, the deposited biological resources belong to the State and the Designated Repository acts as their custodian mandated for preserving/conserving them for five years or beyond, as required by the NBA, under appropriate storage conditions from the date of their deposition. Further, if any new taxon is deposited, it should be preserved for indefinite period. During this period, the deposited biological resources will be made available for research/academic purposes within the country. However, researchers covered under the category of applicants, covered u/s 3 (2) of the Act, shall be required to obtain prior approval from NBA for accessing the deposited materials.

Sr. No.	Category of Biological Resource	Name of the Designated Repository
1.	Flora (angiosperms, gymnosperms, pteridophytes, bryophytes, lichens, macro fungi, macro algae)	Botanical Survey of India and its Regional centres. Indian Council of Forestry Research (FRI, Dehradun and IFGTB, Coimbatore; NBRI, Lucknow
2.	Marine Flora and Fauna	NIO, Goa
3.	Microorganisms	IMTECH, Chandigarh (actinobacteria, bacteria, fungi and yeasts) NCCS, Pune (bacteria, fungi (including yeasts), recombinant DNA materials (in the form of clones in bacterial host) and bacteriophages) IARI, New Delhi (fungi / blue-green algae)
4.	Viruses	NIV, Pune
5.	Genetic Resources	NBAIM Agriculturally important Microorganisms

8. Procedure for Submission of Biological Resources to DRs:

All DRs will develop detailed standard procedures depending upon the type of biological resources deposited with them. In accordance with these standard procedures, all applicants/depositors will be required to deposit the biological resources in the relevant DR along with passport data/information as prescribed in Annexure I. This may be modified by the concerned DR, as required for the kind of biological resource to be deposited, with prior approval of NBA.

Details of such procedure to deposit the biological resource would be displayed at official website of every DR and it is to be linked with NBA website.

9. Issue of Acknowledgement/Receipt:

On depositing the voucher specimen in appropriate form, all applicants will be issued an acknowledgement (Annexure II) by the Scientist-Incharge of DR in reasonable period after satisfactory deposition of voucher specimen/reference sample.

10. Monitoring:

The Director/Head of the organization where DR is located will be the Competent Authority to monitor the maintenance of the deposited biological resources from time to time. These periodic reports will be sent by him to the NBA Secretariat for information and follow up.

11. Resource Support to DRs on Request:

Depending upon the workload of conserving the deposited materials and maintaining their database, contractual technical manpower may be provided to a DR, where justified for this purpose with appropriate financial support by NBA, if necessary.

Think Over It

1. How do the bacteria eat?
2. What is the basic difference between chemotrophs and phototrophs?
3. In what way would an anaerobic chamber resemble the space laboratory orbiting in the vacuum of space.
4. Whether a colony formed in streak plate method always derived from a single bacterium?

SUMMARY

- Microorganisms require some elements in large quantities to construct carbohydrates, lipids, proteins and nucleic acids (basic cell components). Other elements are required in very small amounts and are part of enzymes and cofactors.

- All microorganisms are placed in one of a few nutritional categories based on their requirements for carbon, energy, and hydrogen atoms or electrons.

- Culture media are necessary to grow microorganisms in the laboratory. They are divided in to two categories, synthetic and non-synthetic.

- Pure culture consists of a nutrient medium containing the growth of a single species of organism. Pure cultures can be obtained by using spread plate, streak plate or pour plate method.

- Phototrophs are organisms which depend on light as a source of energy while Chemotrophs are organisms that use chemical compounds.

- Autotrophs are the organisms which use CO_2 (inorganic source) as principal carbon source while Heterotrophs are the organisms which obtain carbon from organic nutrients source of energy.

- Some of the environmental parameters which are related to the culture medium are pH, temperature, oxygen concentration and provision of light.

- Common ingredients of media include peptone, yeast extract, meat extract, sodium chloride, mineral salts, agar and water.

- Based on nature of ingredients, there are two types of media, living media and non-living media. Non-living media includes Natural media, Semisynthetic media and Synthetic media.

- Based on function or application, the media includes selective media, enriched media, enrichment media, differential media.

- Based on environmental conditions extremophiles are classified as thermophiles, psychrophiles, alkaliphiles, acidophiles, barophiles, halophiles, radoduric and metallotolerant.

- Cultivation of anaerobes is carried out by use of special anaerobic culture media, anaerobic chamber, anaerobic bags or pouches and anaerobic jars.

- For the cultivation of viruses living cells are required.

- For the maintenance of bacterial and fungal pure cultures the methods used include periodic transfer to fresh media, refrigeration, paraffin method, soil/grain cultures, cryopreservation and lyophilization.

- The role of culture collection centres is to serve as national facility for microbial culture collection, deposition, maintenance of important and useful microbial cultures and supply of the cultures on demand, and preparation of their informative documents.

Exercise

(A) Questions for One Mark:

1. What are chemoautotrophs / chemoheterotrophs / photoautotrophs / photoheterotrophs?
2. Give any two examples of
   (i) Chemoautotrophs
   (ii) Chemoheterotrophs
   (iii) Photoautotrophs
   (iv) Photoheterotrophs
   (v) Selective media
   (vi) Differential media
   (vii) Methods for preservation of microbial cultures
   (viii) Halophiles
   (ix) Thermophiles
   (x) Psychrophiles
3. What is a growth medium?
4. Bacteria which tolerate high salt concentration are called as
   a) Barophiles   b) Mesophiles
   c) Halophiles   d) None of these
5. PHB granules are used for storage of
   a) Carbohydrates   b) Sulphur
   c) Lipids   d) Phosphate
6. The organism which obtain their energy from chemicals are designated as
   a) Prototrophs   b) Chemotrophs
   c) Organotrophs   d) Autotrophs

(B) Questions for Two Marks:

1. How the microorganisms are divided based on their oxygen requirement?
2. Define the following
   (i) Pure Culture

    (ii)   Selective media

    (iii) Complex media

    (iv) Enrichment media

    (v)   Differential media

3. Differentiate between: selective and differential medium.

4. How the pH of medium affects microbial growth?

(C) Questions for Four Marks:

1. How are protozoa/algae cultivated in the laboratory?

2. Illustrate with the help of flowchart the classification of microorganisms based on their nutritional requirements.

3. Explain the principles on which the growth medium is constructed?

4. Justify:

    (i) MacConkey's medium is both selective and differential.

    (ii) Enrichment is required for the isolation of some organisms.

5. Enlist the methods used for preservation of microbial cultures. Elaborate on any two methods.

6. Compare: (i)   Chemotrophs and phototrophs

           (ii)   Synthetic and non-synthetic medium

           (iii) Enriched and enrichment medium

7. Explain with the help of two examples, the special conditions required for cultivation of extremophiles.

(D) Write short notes on (Four marks each)

    (i)   Lyophilization

    (ii)   Temperature of incubation for microbial growth

    (iii) Solidifying agent in the medium

    (iv) Growth factors requirement of microorganisms

    (v)   Selective media

    (vi) Differential media

    (vii) Extremophiles

    (viii) Streak plate method/Pour plate method/ Spread plate method

    (ix) Nutritional requirements of protozoa

Chapter 2...

Bacterial Growth

Learning Objectives...

➢ To describe a typical growth curves of bacterial population and reasons for lag phase, exponential phase, stationary phase and death phase.

➢ To understand the effect of temperature, pH, nutrient availability and oxygen availability on growth.

Jacques Lucien Monod

Jacques Lucien Monod (February 9, 1910 – May 31, 1976) was a French biochemist who won the Nobel Prize in Physiology or Medicine in 1965, sharing it with François Jacob and André Lwoff "for their discoveries concerning genetic control of enzyme and virus synthesis". Monod and Jacob became famous for their work on the *E. coli* lac operon, which encode proteins necessary for the transport and breakdown of the sugar lactose (lac). From their own work and the work of others, they came up with a model for how the levels of some proteins in a cell are controlled.

Study of the control of expression of genes in the lac operon provided the first example of a system for the regulation of transcription. Monod also suggested the existence of messenger RNA molecules that link the information encoded in DNA and proteins. For these contributions he is widely regarded as one of the founders of molecular biology.

2.1 INTRODUCTION

Bacterial growth is the division of one piece of bacteria into two daughter cells in a process called binary fission. Assuming that no mutation occurs the resulting daughter cells are genetically identical to the original cell. The doubling of the bacterial population occurs. Both daughter cells from the division do not necessarily survive.

Cell growth refers to the increase in its mass and physical size controlled by physical, biological and chemical environments. Microbial growth is quantified by increase in the macromolecular and chemical constituents of the cell. The growth pattern of each microbe is unique. Cell growth and cell division are inseparable for microbes as bacteria divided by binary fission.

2.2 LOGISTIC GROWTH

Logistic population growth occurs when the growth rate decreases as the population reaches carrying capacity. Carrying capacity is the maximum number of individuals in a population that the environment can support. A graph of logistic growth is shaped like an S.

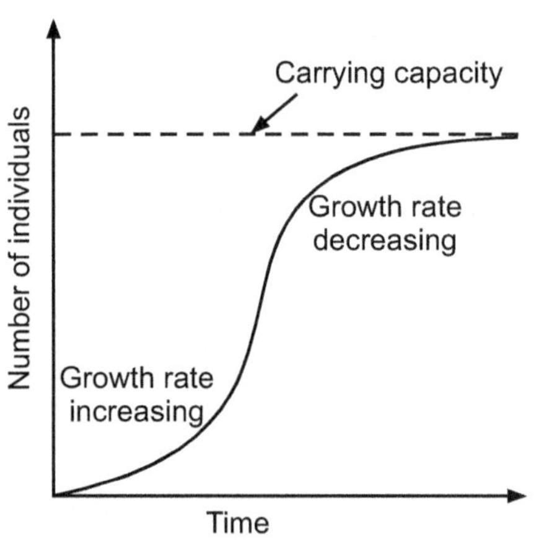

Fig. 2.1

Exponential population growth: When resources are unlimited, populations exhibit exponential growth, resulting in a J-shaped curve. Growth kinetics is an autocatalytic reaction which implies that the rate of growth is directly proportional to the concentration of cell.

Microbial growth kinetics, i.e., the relationship between the specific growth rate (μ) of a microbial population and the substrate concentration (s), is an indispensable tool in all fields of microbiology.

The Monod equation is a mathematical model for the growth of microorganisms. It is named for Jacques Monod who proposed using an equation of this form to relate microbial growth rates in an aqueous environment to the concentration of a limiting nutrient.

The Monod equation is:

$$\mu = \mu_{max}\, S/(K_S + S)$$

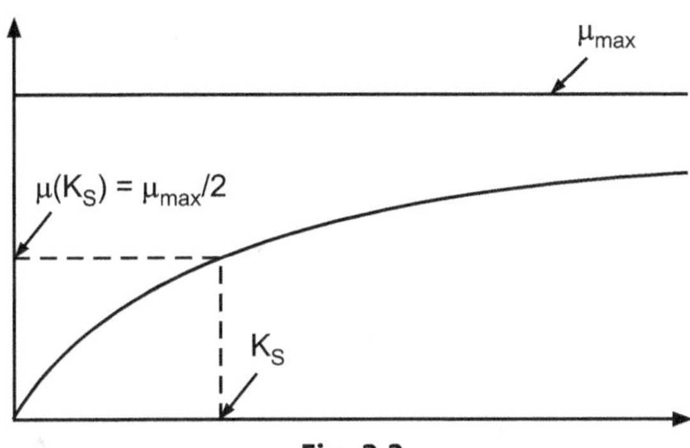

Fig. 2.2

The specific growth rate μ as a function of substrate concentration S.

where,

- μ is the specific growth rate of the microorganisms
- μ_{max} is the maximum specific growth rate of the micro-organisms
- S is the concentration of the limiting substrate for growth
- K_S is the "half-velocity constant", the value of S when μ/μ_{max} = 0.5

μ_{max} and K_S are empirical coefficients to the Monod equation. They will differ between species and based on the ambient environmental conditions

Cell increase rate differential equation:

$$dX/dt = \mu X$$

Solution: $X = X_o\, e^{\mu t}$

where,

X = Cell number or concentration

μ = Specific growth rate

X_o = Cell number at t = 0

[With growth requirements in excess]

$\mu = \mu_m$ and is characteristic of a particular bacterial species

Our substrate disappearance rate can be described as:

$$U = 1/X\ dS/dt$$

where,

U = Specific substrate utilization rate

S = Substrate concentration

Substrate concentration and cell growth are interrelated. As cells grow, they use the substrate so it disappears. We can write the expression:

$$dX/st = -Y \, dS/dt$$

where, the right-hand side is negative to denote disappearance as the left-hand side increases, Y is a yield factor (dimensionless or rather in mass cells/mass substrate).

Not all substrate goes into cells for synthesis, some must go for energy end products of CO_2, H_2O or CH_4.

Therefore, $Y < 1$

But, we know in reality, a substrate decreases and we get a growth cycle.

Several kinetic models have been developed to describe this type of substrate-based growth.

Logistic Growth Model:

Used from beginning of exponential phase through end of max population (or stationary phase)

Modification of unlimited growth eqn

$$dX/dt = \mu_m \, X[1 - X/X_f]$$

(Terms same as before)

X_f = Maximum population size or concentration

Term $[1 - X/X_f]$ describes reduction of specific growth rate as exponential growth phase gives way to retardation phase.

Monod Growth Model:

$$\mu = \mu_m S/(K_S + S)$$

All terms described previously except K_S = half saturation constant or substrate concentration S, at which $\mu = 1/2\mu_{max}$.

Combine above eqns:

$$dS/dt = -\mu_m/Y \, SX/(K_S + S)$$
$$\mu_m/Y \text{ replaced with K}$$
$$-dS/dt = KSX/(K_S + S)$$

Bacterial growth in batch culture can be presented with four different phases: lag phase (A), log phase or exponential phase (B), stationary phase (C), and death phase (D).

During lag phase, bacteria adapt themselves to growth conditions. It is the period where the individual bacterial cells are attaining maturity and not yet able to divide. During this phase of the bacterial growth cycle, synthesis of RNA, enzymes and other molecules occurs. Length of this phase depends upon nature of the medium, size of inoculums, species of organism and various physical and chemical factors. If an exponentially growing culture is inoculated into identical medium which is kept at same temperature a very small or no lag phase will be observed.

The log phase (sometimes called the logarithmic phase or the exponential phase) is a period characterized by cell doubling. The number of new bacteria appearing per unit time is proportional to the present population. If growth is not limited, doubling will continue at a constant rate so both the number of cells and the rate of population increase doubles with each consecutive time period. For this type of exponential growth, plotting the natural logarithm of cell number against time produces a straight line. The slope of this line is the specific growth rate of the organism, which is a measure of the number of divisions per cell per unit time. The actual rate of this growth (i.e. the slope of the line in the figure) depends upon the growth conditions, affecting the frequency of cell division events and the probability of both daughter cells surviving. As the cells are constantly dividing they are small in size. The entire population is uniform in terms of chemical composition, metabolic activities and other physiological properties. This is the reason why log phase cultures are used in biochemical and physiological studies. As the cells are young and small in this phase they are more sensitive to various physical and chemical agents.

Because the medium is soon depleted of nutrients and enriched with wastes exponential growth cannot continue indefinitely.

The stationary phase is often due to a growth-limiting factor such as the depletion of an essential nutrient, and/or the formation of an inhibitory product such as an organic acid. Stationary phase results from a situation in which growth rate and death rate are equal. The number of new cells created is limited by the growth factor and as a result the rate of cell growth matches the rate of cell death. The result is a "smooth," horizontal linear part of the curve during the stationary phase.

At death phase, bacteria run out of nutrients and die. The rate of cell death is maximum and constant in this phase. Number of cells decreases exponentially with time. If log number of cells is plotted against time a straight line with negative slope is obtained. Death of cells occurs due to many factors such as exhaustion of nutrients, accumulation of toxic metabolites (waste) and the changes in ion equilibrium especially pH. Death rate depends upon type of species and environmental conditions.

Batch culture is the most common laboratory growth method in which bacterial growth is studied, but it is only one of the many methods. The bacterial culture is incubated in a closed vessel with a single batch of medium. In some experimental regimes, some of the bacterial culture is periodically removed and added to fresh sterile medium. In the extreme case, this leads to the continual renewal of the nutrients. This is a chemostat, also known as continuous culture. It is, in a steady state defined by the rates of nutrient supply and bacterial growth. In comparison to batch culture, bacteria are maintained in exponential growth phase, and the growth rate of the bacteria is known. Related devices include turbidostats and auxostats.

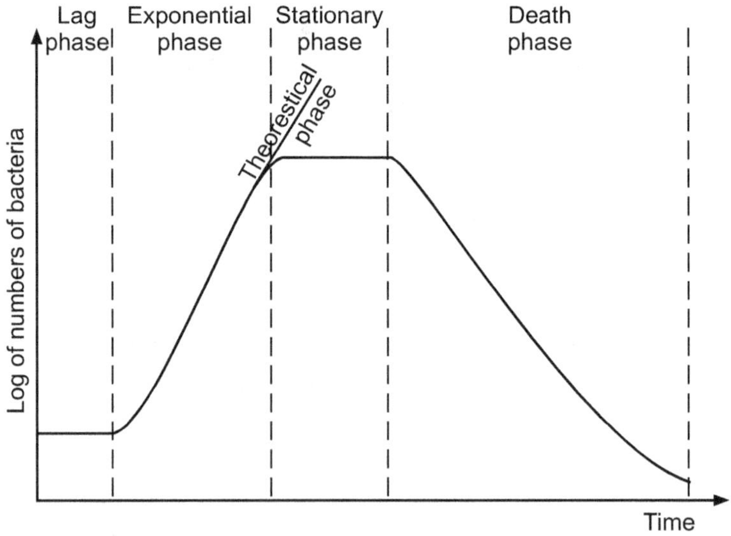

Fig. 2.3

Bacterial growth can be suppressed with bacteriostats, without necessarily killing the bacteria. In a near natural situation in which more than one bacterial species is present, the growth of microbes is more dynamic and continual.

Liquid is not the only laboratory environment for bacterial growth. Spatially structured environments such as biofilms or agar surfaces present additional complex growth models.

2.3 GENERATION TIME

The time taken by the bacteria to double in number during a specified time period is known as the generation time. The generation time tends to vary with different organisms. *E. coli* divides in every 20 minutes, hence its generation time is 20 minutes, and for *Staphylococcus aureus* it is 30 minutes.

Generation time (G) is defined as the time (t) per generation (n = number of generations). Hence, G = t/n is the equation from which calculations of generation time derive.

The exponential phase (sometimes called the log phase or the logarithmic phase) is a period characterized by cell doubling. The number of new bacteria appearing per unit time is proportional to the present population.

Table 2.1: Generation time for some bacteria under optimum conditions

Bacterium	Medium	Temperature (°C)	Generation time (min.)
Bacillus thermophilus	Broth	55	1.3
Escherichia coli	Broth	37	20
Bacillus subtilis	Broth	37	27
Streptococcus lactis	Milk	37	30
Lactobacillus acidophilus	Milk	37	66 – 87
Mycobacterium tuberculosis	Synthetic	37	792 – 932
Treponema pallidium	Rabbit testes	37	1980

2.4 DIAUXIC GROWTH

A diauxic growth curve refers to the growth curve generated by an organism which has two growth peaks. Diauxic growth, meaning double growth, is caused by the presence of two sugars on a culture growth media, one of which is easier for the target bacterium to metabolize. The preferred sugar is consumed first, which leads to rapid growth, followed by a lag phase. During this lag phase the cellular machinery used to metabolize the second sugar is activated and subsequently the second sugar is metabolized. The first phase is the fast growth phase, since the bacterium is consuming (in the case when the two sugars are glucose and lactose) exclusively glucose, and is capable of rapid growth. The second phase is a lag phase while the genes used in lactose metabolism are expressed and observable cell growth stops. This is followed by another growth phase which is slower than the first because of the use of lactose as the primary energy source. The final stage is the saturation phase.

The cause of diauxic (diphasic) growth is complex and not completely understood, it is considered that catabolite repression (The genes responsible for the utilization of the complex sugar of the two are repressed till the simple sugar is utilized) or the glucose effect probably plays a part in it.

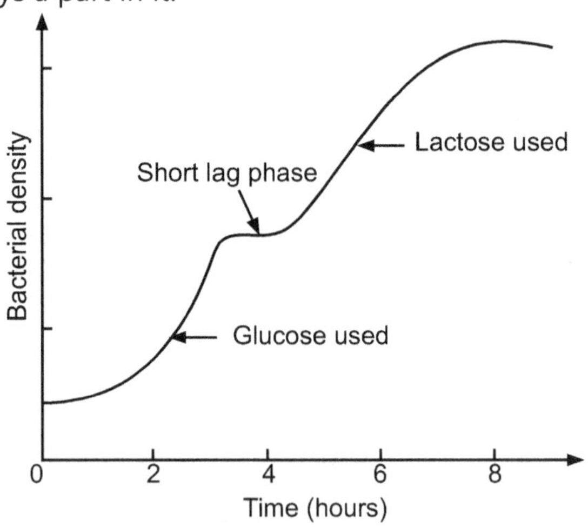

Fig. 2.4: Diauxic or diphasic growth curve of *E. coli* grown with a mixture of glucose and lactose. Glucose is first used, then lactose. A short lag phase in diauxic growth is present during which the bacterium synthesizes the enzymes needed for use of lactose

The methods of enumeration in microbes can be divided into four categories. Direct methods involve counting the microbes, while indirect methods involve estimation. Viable methods only count cells that are metabolically active, while total counts include dead and inactive cells.

2.5 DIRECT MICROSCOPIC METHOD (TOTAL CELL COUNT)

Direct microscopic counts are performed by spreading a measured volume of sample over a known area of a slide, counting representative microscopic fields, and relating the averages back to the appropriate volume-area factors. The known volume of sample is smeared uniformly on glass slide in an area marked 1 cm^2. Organisms are stained using any suitable stain and observed under OEL.

Then area of microscopic field is calculated from which total number of fields in 1 cm^2 area can be calculated. Number of organisms per field is determined from which number of organisms per ml of given sample can be calculated.

The formula used for the direct microscopic count is:

The number of bacteria per cc = The average number of bacteria per large double-lined square × The dilution factor of the large square (1,250,000) × The dilution factor of any dilutions made prior to placing the sample in the counting chamber, e.g., mixing the bacteria with dye.

2.5.1 Petroff-Hauser Chamber

Specially constructed counting chambers, such as the Petroff-Hauser and Levy counting chambers, simplify the direct counting procedure because they are made with depressions in which a known volume overlies an area that is ruled into squares. The ability to count a defined area and convert the numbers observed directly to volume makes the direct enumeration procedure relatively easy. Direct counting procedures are rapid but have the disadvantage that they do not discriminate between living and dead cells.

The Petroff-Hausser counting chamber has small etched squares 1/20 of a millimeter (mm) by 1/20 of a mm and is 1/50 of a mm deep. The volume of one small square therefore is 1/20,000 of a cubic mm or 1/20,000,000 of a cubic centimeter (cc). There are 16 small squares in the large double-lined squares that are actually counted, making the volume of a large double-lined square 1/1,250,000 cc. The normal procedure is to count the number of bacteria in five large double-lined squares and divide by five to get the average number of bacteria per large square. This number is then multiplied by 1,250,000 since the square holds a volume of 1/1,250,000 cc, to find the total number of organisms per cc in the original sample.

2.5.2 Improved Neubauer Chamber

The Neubauer chamber is designed to leave a gap of 100 mm between the top surface of the counting area and the bottom surface of the cover glass. The improved Neubauer has a slightly different grid pattern compared to the old Neubauer chamber.

The gridded area of the hemocytometer consists of nine 1×1 mm (1 mm^2) squares. These are subdivided in three directions; 0.25×0.25 mm (0.0625 mm^2), 0.25×0.20 mm (0.05 mm^2) and 0.20×0.20 mm (0.04 mm^2). The central square is further subdivided into 0.05×0.05 mm (0.0025 mm^2) squares. The raised edges of the hemocytometer hold the cover slip 0.1 mm off the marked grid, giving each square a defined volume.

Table 2.2

Dimensions	Area	Volume at 0.1 mm depth
1×1 mm	1 mm^2	100 nL
0.25×0.25 mm	0.0625 mm^2	6.25 nL
0.25×0.20 mm	0.05 mm^2	5 nL
0.20×0.20 mm	0.04 mm^2	4 nL
0.05×0.05 mm	0.0025 mm^2	0.25 nL

To use the hemocytometer, first make sure that the special coverslip provided with the counting chamber is properly positioned on the surface of the counting chamber. When the two glass surfaces are in proper contact, Newton's rings can be observed. If so, the cell suspension is applied to the edge of the coverslip to be sucked into the void by capillary action which completely fills the chamber with the sample. The number of cells in the chamber can be determined by direct counting using a microscope, and visually distinguishable cells can be differentially counted. The number of cells in the chamber is used to calculate the concentration or density of the cells in the mixture the sample comes from. It is the number of cells in the chamber divided by the chamber's volume, which is known from the start, taking account of any dilutions and counting shortcuts:

Particles per ml volume

$$= \frac{\text{Counted particles}}{\text{Counted surface (mm}^2) \times \text{Chamber depth (mm)} \times \text{Dilution}}$$

Where the volume of the diluted sample (after dilution) divided by the volume of the original mixture in the sample (before dilution) is the dilution factor. For example, if the volume of the original mixture was 20 µL and it was diluted once (by adding 20 µL dilant), then the second term in parentheses is 40 µL/20 µL. The volume of the squares counted is the one shown in the table 2.2, depending on the size. The number of cells counted is the sum of all cells counted across squares in one chamber. The proportion of the cells counted applies if not all inner squares within a set square are counted (i.e., if only 4 out of the 20 in a corner square are counted, then this term will equal to 0.2).

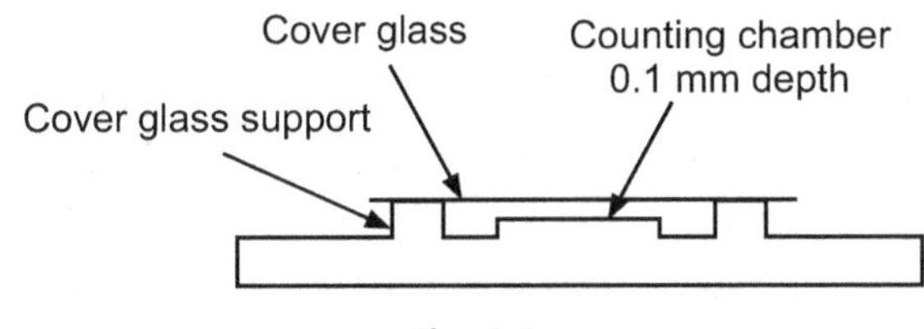

Fig. 2.5

The parts of the hemocytometer are identified. For most applications, the four large corner squares are only used. The cells that are on or touching the top and left lines are counted, but the ones on or touching the right or bottom lines are ignored.

2.6 TOTAL VIABLE COUNT

The number of bacteria in a given sample is usually too great to be counted directly. However, if the sample is serially diluted and then plated out on an agar surface in such a manner that single isolated bacteria form visible isolated colonies, the number of colonies can be used as a measure of the number of viable (living) cells in that known dilution. However, keep in mind that if the organism normally forms multiple cell arrangements, such as chains, the colony-forming unit may consist of a chain of bacteria rather than a single bacterium. In addition, some of the bacteria may be clumped together. Therefore, when doing the plate count technique, generally we are determining the number of Colony-Forming Units (CFUs) in that known dilution. By extrapolation, this number can in turn be used to calculate the number of CFUs in the original sample.

Normally, the bacterial sample is diluted by factors of 10 and plated on agar. After incubation, the number of colonies on a dilution plate showing between 30 and 300 colonies is determined. A plate having 30-300 colonies is chosen because this range is considered statistically significant. If there are less than 30 colonies on the plate, small errors in dilution technique or the presence of a few contaminants will have a drastic effect on the final count. Likewise, if there are more than 300 colonies on the plate, there will be poor isolation and colonies will have grown together.

Generally, one wants to determine the number of CFUs per milliliter (ml) of sample. To find this, the number of colonies (on a plate having 30-300 colonies) is multiplied by the number of times the original ml of bacteria is diluted (the dilution factor of the plate counted). For example, if a plate containing a 1/1,000,000 dilution of the original ml of sample shows 150 colonies, then 150 represents 1/1,000,000 the number of CFUs present in the original ml. Therefore the number of CFUs per ml in the original sample is found by multiplying 150 × 1,000,000 as shown in the formula below:

The number of CFUs per ml of sample = The number of colonies (30-300 plate) × The dilution factor of the plate counted

For a more accurate count it is advisable to plate each dilution in duplicate or triplicate and then find an average count.

2.7 TURBIDIMETRIC DETERMINATION

Turbidimetric determination is useful for plotting growth curves of bacteria in broth or liquid media. It is one of the simplest methods used to analyze trends in growth because it uses a spectrophotometer to track changes in the optical density (OD) over time.

A quick and efficient method of estimating the number of bacteria in a liquid medium is to measure the turbidity or cloudiness of a culture and translate this measurement into cell numbers. This method of enumeration is fast and is usually preferred when a large number of cultures are to be counted. 34 Although measuring turbidity is much faster than the standard plate count, the measurements must be correlated initially with cell number. This is achieved by determining the turbidity of different concentrations of a given species of microorganism in a particular medium and then utilizing the standard plate count to determine the number of viable organisms per milliliter of sample. A standard curve can then be drawn in which a specific turbidity or optical density reading is matched to a specific number of viable organisms. Subsequently, only turbidity needs to be measured. The number of viable organisms may be read directly from the standard curve, without necessitating time-consuming standard counts. Turbidity can be measured by an instrument such as a colorimeter or spectrophotometer. These instruments contain a light source and a light detector (photocell) separated by the sample compartment. Turbid solutions such as cell cultures interfere with light passage through the sample, so that less light hits the photocell than would if the cells were not there. Turbidimetric methods can be used as long as each individual cell blocks or intercepts light; as soon as the mass of cells becomes so large that some cells effectively shield other cells from the light, the measurement is no longer accurate. Before turbidimetric measurements can be made, the spectrophotometer must be adjusted to 100% transmittance (0% absorbance). This is done using a sample of uninoculated medium. Percent transmittance oroptical density of various dilutions of the bacterial culture is then measured. A wavelength of 420 nm is used when the solution is clear, 540 nm when the solution is light yellow, and 600-625 nm is used for yellow to brown solutions.

2.8 DETERMINATION OF DRY WEIGHT

This is one of the simplest indirect methods in situations where determining the number of microorganisms is difficult or undesirable for other reasons. These methods measure some quantifiable cell property that increases as a direct result of microbial growth. The simplest technique of this sort is to measure the weight of cells in a sample. Portions of a culture can be taken at particular intervals and centrifuged at high speed to sediment bacterial cells to the bottom of a vessel. The sedimented cells (called a cell pellet) are then washed to remove contaminating salt, and dried in an oven at 100-105°C to remove all water, leaving only the mass of components that make up the population of cells. An increase in the dry weight of the cells correlates closely with cell growth. However, this method will count dead as well as living cells. There might also be conditions where the dry weight per cell changes over time or under different conditions. For example, some bacteria that excrete polysaccharides will have a much higher dry weight per cell when growing on high sugar levels (when polysaccharides are produced) than on low. If the species under study forms large clumps of cells such as those that grow filamentously, dry weight is a better measurement of the cell population than is a viable plate count.

2.9 PACKED CELL VOLUME

For some types of work the volume occupied by organisms has proven a satisfactory measure of the number of cells. Total cell volume is determined by centrifuging, the sediment being deposited in a calibrated tube from which the volume of packed cells is read directly. Having already established the volume-cell number ratio, preferably by direct count, volume readings are readily convertible into total cells per cubic centimeter. The centrifuge tubes (Haematocrit tubes) constitute the essential feature of the method. They are calibrated individually with dense cell suspensions which are made up in buffer solution, and standardized by direct count. Five consecutive 1 ml aliquots are centrifuged at known speed for a period of time sufficient to insure maximal packing. The column length of the sediment is measured in each instance and the average bacterial content per millimeter of tube length calculated. Both menisci are level and sharp and permit accurate measurement. The method is suitable for work requiring rapid and accurate routine preparation of bacterial suspensions of known cell concentrations.

2.10 CHEMICAL METHODS (CELL CARBON AND NITROGEN ESTIMATION)

2.10.1 Estimation of Cell Carbon

Estimates of bacterial biomass (as carbon content) are essential to determine the quantitative importance of bacteria. No reliable method for the direct determination of bacterial biomass is currently available. Estimates can be made, however, by converting bacterial bio volume into organic carbon. The conversion factor involved may be calculated from values for the buoyant density, the dry weight/wet weight ratio, and the carbon weight/dry weight ratio of bacterial cells. Luria (13) states that about 20% is a reasonable average estimate of bacterial dry weight. Percent dry matter of bacterial cells is usually determined by weighing a bacterial pellet before and after drying to constant weight.

Studies have estimated bacterial biomass based on the assumption that one bacterial cell contains 20 fg of carbon.

The bacterial community is usually cultivated until its growth reaches a plateau phase. Then, the cell culture is concentrated by filtering through glass fiber filter (GF/F), and then the carbon content is determined by one of the methods like particulate organic carbon retained on a glass fiber filter. The limitations of the method includes (1) the prefiltration, which excludes large bacteria from the filtrate; (2) the longterm incubation of the isolated bacterial community on different media or with growth stimulators; (3) using of GF/F filters, which miss an essential part of small bacterial cells. Since the methods of bacteria preparation for the analyses are quite labor intensive and time consuming and since permanent longterm observations are needed to obtain the cell culture in an appropriate growth phase, an empirically or experimentally determined "volume : biomass (dry weight or carbon content)" conversion factor is used, which is determined by experimental methods. When processing routine samples, it is usually assumed that a bacterial cell contains about 80–85% water out of the wet weight and 50% carbon out of the dry weight.

To convert bacterial biovolume into biomass (carbon content), its suggested that 0.22 g of C /cm^3 should be used as a conversion factor.

2.10.2 Estimation of Cell Nitrogen

The major constituent of cell material is protein, and since nitrogen is a characteristic part of proteins, one can measure a bacterial population or cell crop in terms of bacterial nitrogen. Bacteria average approximately 14 % nitrogen on a dry weight basis, although this figure is subject to some variation introduced by changes in culture conditions or differences between species. To measure growth by this technique, you must first harvest the cells and wash them free of medium and then perform a quantitative chemical analysis for nitrogen by microkjeldahl method. Bacterial nitrogen determinations are somewhat laborious and can be performed only on specimens free of all other sources of nitrogen. Furthermore, the method is applicable only for concentrated populations. For these and other reasons, this procedure is used primarily in research.

2.11 FACTORS AFFECTING BACTERIAL GROWTH

2.11.1 Effects of pH on Bacterial Growth

Acidity is a function of the concentration of hydrogen ions [H$^+$] and is measured as pH. Environments with pH values below 7.0 are acidic, with a high concentration of H$^+$ ions; those with pH values above 7.0 are considered basic. Extreme pH affects the structure of all macromolecules. The hydrogen bonds holding together strands of DNA break up at high pH. Lipids are hydrolyzed by an extremely basic pH. The proton motive force responsible for production of ATP in cellular respiration depends on the concentration gradient of H$^+$ across the plasma membrane. If H$^+$ ions are neutralized by hydroxide ions, the concentration gradient collapses and impairs energy production. But the component most sensitive to pH in the cell is the protein. Moderate changes in pH modify the ionization of amino-acid functional groups and disrupt hydrogen bonding, which, in turn, promotes changes in the folding of the molecule, promoting denaturation and destroying activity.

The optimum growth pH is the most favourable pH for the growth of an organism. The lowest pH value that an organism can tolerate is called the minimum growth pH and the highest pH is the maximum growth pH.

Most bacteria are neutrophiles, meaning they grow optimally at a pH within one or two pH units of the neutral pH of 7, between 5 and 8. Most familiar bacteria, like *Escherichia coli, staphylococci,* and *Salmonella spp.* are neutrophiles and does not fare well in the acidic pH Fungi thrive at slightly acidic pH values of 5.0 – 6.0.

Microorganisms that grow optimally at a pH less than 5 are called acidophiles. For example, the sulphur-oxidizing Sulfolobus spp. isolated from sulphur mud fields and hot springs are extreme acidophiles. Lactobacillus bacteria, which are an important part of the normal microbiota of the vagina, can tolerate acidic environments at pH values 3.5 – 6.8.

Acidophilic microorganisms display a number of adaptations to survive in strong acidic environments. While the membrane is slightly leaky to protons, the cytoplasmic pH of most acidophiles is generally only slightly acidic. One of the major reasons for this is their ability to actively transport of H^+ ions out of the cell. In addition, cytoplasmic proteins have evolved to function better at a slightly acidic pH with increased negative surface charges compared to their neutrophiles.

The alkaliphiles are microorganisms that have pH optima between 8.0 and 11. Vibrio cholerae, the pathogenic agent of cholera, grows best at the slightly basic pH of 8.0; it can survive pH values of 11.0 but is inactivated by the acid. Extreme alkaliphiles have adapted to their harsh environment through various evolutionary modifications.

Microorganisms grow best at their optimum pH. Growth occurs slowly or not at all below the minimum growth pH or not at all below the minimum growth pH and above the maximum growth pH.

2.11.2 Effect of Temperature on Bacterial Growth

An increase in temperature will increase enzyme activity. But if temperatures get too high, enzyme activity will diminish and the protein (the enzyme) will denature.

On the other hand, lowering temperature will decrease enzyme activity. At freezing temperatures enzyme activity can stop. Repeated cycles of freezing and thawing can denature proteins. In addition, freezing causes water to expand and also forms ice crystals, hence cells begin to rupture.

Every bacterial species has specific growth temperature requirements which are largely determined by the temperature requirements of its enzymes. Each organism will have:

- a minimum growth temperature,
- an optimum growth temperature,
- a maximum growth temperature.

Organisms can be classified according to their optimum growth temperature.

- Psychrophiles grow best between –5°C and 20°C.
- Mesophiles grow best between 20°C and 45°C and
- Thermophiles grow best at temperatures above 45°C.
- Thermoduric organisms can survive high temperatures but don't grow well at such temperatures. Organisms which form endospores would be considered thermoduric.

2.11.3 Effect of Solute Concentration on Bacterial Growth

1. Salt:

For most bacteria, increasing osmolarity decreases bacterial growth rate. Some bacteria have a preferred osmolarity for maximum growth.

Most bacteria do not need to regulate their internal osmolarity with precision because their cell walls can sustain a considerable osmotic pressure.

The availability of water is the critical factor that affects the growth of cells.

The availability of water is measured by the water activity, A_w. For pure water, A_w is 1.00.

As the osmolarity increases, the Aw decreases. Some typical A_w values are

- Blood, 0.99
- Seawater, 0.99
- Maple syrup, 0.90
- Great Salt Lake 0.75

The most common solute in nature is salt, NaCl.

We can classify bacteria according to their growth response to salt. Non-halophiles such as *E. coli*, Pseudomonas and Aquaspirillum serpens cannot tolerate even low levels of salt concentration.

Halotolerant bacteria such as *Staphylococcus aureus* grow best in the absence of NaCl but can tolerate moderate levels of salt concentration.

Mild Halophiles such as Alteromonas haloplanktis and *Pseudomonas marina* can tolerate 1-6% salt (sea water is 3% salt).

Moderate Halophiles such as *Paracoccus halodenitrificans* and *Vibrio costicolus* grow best in a medium that contains 6-15% salt.

Extreme Halophiles such as *Halococcus morrhuae, Halobacterium salinarium,* and *Pediococcus halophilus* thrive in a medium that contains 15-36% salt.

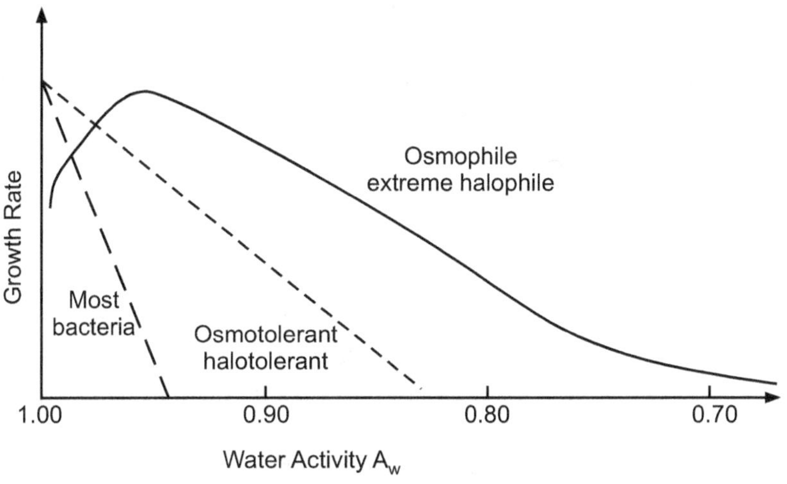

Fig. 2.6

From left to right, the graph above shows the growth rate of a non-halophile, a halotolerant bacterium, and an extreme halophile.

2. Sugar:

There are several ways in which sugar inhibit microbial growth. The most notable is simple osmosis, or dehydration. Sugar, whether in solid or aqueous form, attempts to reach equilibrium with the salt or sugar content of the food product with which it is in contact. This has the effect of drawing available water from within the food to the outside and inserting sugar molecules into the food interior. The result is a reduction of the water activity (a_w), a measure of unbound, free water molecules in the food that is necessary for microbial survival and growth. The a_w of most fresh foods is 0.99 whereas the a_w necessary to inhibit growth of most bacteria is roughly 0.91.

Other antimicrobial mechanisms of sugar include interference with a microbe's enzyme activity and weakening the molecular structure of its DNA. Sugar may also provide an indirect form of preservation by serving to accelerate accumulation of antimicrobial compounds from the growth of certain other organisms. Examples include the conversion of sugar to ethanol in wine by fermentative yeasts or the conversion of sugar to organic acids in sauerkraut by lactic acid bacteria.

Microorganisms differ widely in their ability to resist sugar-induced reductions of a_w. Most disease-causing bacteria do not grow below 0.94 a_w, whereas most molds that spoil foods grow at an a_w as low as 0.80, corresponding to highly concentrated sugar solutions. Yet other microorganisms grow quite well under even more highly osmotic, low a_w conditions. Food products that are concentrated sugar solutions, such as concentrated fruit juices, can be spoiled by sugar-loving yeasts such as species of Zygosaccharomyces.

2.11.4 Effect of Heavy Metals on Bacterial Growth

Addition of trace amounts of heavy metals to the environment of microbial cells often stimulates microbial growth. However, higher concentrations result in severe reduction of microbial activity, which is reflected by reduction of the apparent growth rate and increase in lag time. This is called as the oligodynamic effect of heavy metals.

Many metallic elements have been observed to inhibit the growth of bacteria and to inactivate enzymes. Practical application of such activity of metals has been made in the purification of water and in the preservation of tomato juice, cider and hides. This antimicrobial effect is shown by metals such as mercury, silver, copper, lead, zinc, gold, aluminium and other metals, and the concentration of the metal needed for this antimicrobial effect is extremely small. The exact mechanism of this action is still unknown but some data suggest that the metal ions denature protein of the target cells by binding to reactive groups resulting in their precipitation and inactivation. The high affinity of cellular proteins for the metallic ions results in the death of the cells due to cumulative effects of the ion within the cells. Similarly, silver inactivates enzymes by binding with sulfhydryl groups to form silver sulfides or sulfhydryl-binding propensity of silver ion disrupts cell membranes, disables proteins and inhibits enzyme activities. The study also suggest that positively charged copper ion distorts the cell wall by bonding to negatively charged groups and allowing the silver ion into the cell Silver ions bind to DNA, RNA, enzymes and cellular proteins causing cell damage and death. Both gram positive and negative bacteria are affected by the oligodyanmic action of heavy metals. However, there are several bacterial strains that contain genetic determinants of resistance to heavy metal. These determinants for resistance are often found in plasmids and trasposons.

Think Over It

1. Growth curve has four distinct phases.
2. Why does generation time vary with species of bacteria?
3. Is catabolite repression genetically controlled?
4. Are microscopic methods reliable as counting methods for bacteria?
5. Why salts and sugars are used as preservatives?

SUMMARY

- Bacterial growth is the division of one piece of bacteria into two daughter cells.
- Cell growth refers to the increase in its mass and physical size controlled by physical, biological and chemical environments.
- Cell growth and cell division are inseparable for microbes.
- Logistic population growth occurs when the growth rate decreases as the population reaches carrying capacity.
- When resources are unlimited, populations exhibit exponential growth.

- The rate of growth is directly proportional to the concentration of cell.
- The Monod equation is a mathematical model for the growth of microorganisms.
- Bacterial growth can be presented with four different phases: lag phase, log phase or exponential phase, stationary phase, and death phase.
- The time taken by the bacteria to double in number during a specified time period is known as the generation time.
- A diauxic growth curve refers to the growth curve generated by an organism which has two growth peaks.
- There are various methods for measurement of Bacterial growth.
- The microscopic methods for measurement of bacterial growth include direct microscopic count, counting cells using improved Neubauer, Petroff-Hausser's chamber.
- The plate counts include the Total viable count.
- The turbidimetric methods are also a measure of cell growth.
- The biomass can be estimated by measuring dry mass and packed cell volume.
- The chemical methods for estimation of growth include cell carbon and nitrogen estimation.
- The various factors affecting bacterial growth are pH, Temperature, Solute Concentration (Salt and Sugar)} and Heavy metals.

Exercise

(A) Questions for One Mark:

1. Define growth.
2. Define logistic growth.
3. What are the four phases of growth?
4. Define Generation time.
5. State Monod's equation.
6. State methods to count bacteria.
7. What are acidophiles?
8. What are alkaliphiles?

9. What are neutrophiles?
10. What are halophiles?
11. What is oligodynamic effect?
12. Give a few examples of heavy metals.

(B) Fill in the blanks:

1. The portion of the growth curve where a rapid growth of bacteria is observed is known as ___________.
2. The generation time for E.coli is _______
3. Secondary metabolites are produced during --------------
4. --------------is used to grow bacterial culture continuously.
5. --------------refers to the time required for one cell to become two cells.
6. The stage when no bacterial growth occurs is called the------- -----.
7. Bacteria grow and divide by -------------.

(C) Questions for Two Marks:

1. State and explain Monod's equation.
2. Explain log phase of growth.
3. Explain lag phase of growth.
4. Explain stationary phase of growth.
5. Explain death phase of growth.
6. Graphically show phases of growth.
7. Graphically show diauxic growth.
8. State different methods to estimate bacterial growth.
9. Explain acidophiles with examples.
10. Explain alkaliphiles with examples.
11. Explain neutrophiles with examples.
12. Explain halophiles with examples.

(D) Questions for Four Marks:

1. Explain various phases of growth.
2. Explain diauxic growth.
3. Explain DMC.
4. Explain Neubauer Chamber.
5. Explain Petroff-Hausser's chamber.

6. Explain Total viable count.
7. Explain Turbidimetric methods.
8. Explain Dry mass
9.   Explain packed cell volume.
10. Explain cell carbon estimation.
11. Explain cell nitrogen estimation.
12. Explain effect of pH on bacterial growth.
13. Explain effect of salt on bacterial growth.
14. Explain effect of sugar on bacterial growth.
15. Explain effect of temperature on bacterial growth.
16. Explain effect of heavy metals on bacterial growth.

* 9 7 8 9 3 8 9 6 8 6 9 3 7 *